Writing the Engineering Research Article

A Corpus-Based Guide to Structure and Language

Michael Barlow

Athelstan Publications
Suite 464
2368A Rice Blvd
Houston, TX 77005

Contents

Introduction

This book takes a different approach to teaching academic writing. Rather than prescribing rules or offering model paragraphs to imitate, it opens up a large collection of published engineering research articles and asks a simple question: what do writers actually do?

The answer, drawn from a corpus of 163 engineering articles comprising approximately 610,000 words, reveals patterns that no single article could show. When we examine over 150 Introductions, we find that gap-signalling in engineering leans more heavily on technical deficiency, using words and phrases like *challenge* (4.2/10k), *limited* (4.5/10k), *suffer from* (0.9/10k), rather than on research absence, the preferred strategy in other disciplines. When we examine Conclusions, we find that *this paper* at 16.9 per 10,000 words is the dominant way of reintroducing the study at a rate of roughly once every 600 words of Conclusion text. When we look at how results are communicated, we discover that figure references appear at 60.0 per 10,000 words in the technical body sections — roughly once every 17 words, but drop to just 1.5/10k in Conclusions, where the argument shifts from showing to claiming.

These are not rules. They are tendencies: the accumulated choices of over 160 engineering writers navigating the same rhetorical challenges you face. This book makes those tendencies visible so that you can make informed decisions about your own writing.

The Corpus

The analyses in this book draw on a purpose-built corpus of engineering research articles, as described in the table below.

One approach is to classify the articles according to their heading structures: Introduction, Conclusion, and everything in between. This allows us to compare the same linguistic features across different parts of the paper. We can then show, for example, not just that figure references are common in engineering writing, but that they peak in the technical body sections at 60.0 per 10,000 words — roughly 40 times the Conclusion rate of 1.5/10k. Or that *this paper* appears twice as frequently in Conclusions (16.9/10k) as in Introductions (8.7/10k), reflecting the shift from positioning the study to claiming its contribution.

Feature	**Detail**
Articles	163
Total words (body text)	Approximately 610,000
Sections by function	Introduction (~210,000), Technical body sections (~245,000), Methods (~34,000), Results (~16,000), Results+Discussion (~29,000), Discussion (~12,000), Conclusion (~46,000)
Sub-disciplines	Aerospace, General, Mechanical, Civil & Structural, Control & Systems, Electrical & Electronic, Industrial & Manufacturing, Chemical, Materials Science
Total headings analysed	1,623 (617 main headings, 1,006 sub-headings)

Every frequency, every pattern, and every example sentence in this book comes directly from this corpus. The articles span a range of engineering sub-disciplines and methodological approaches — from computational fluid dynamics to structural testing, from control system design to manufacturing processes.

Don't worry if some of the examples are hard to understand; the rhetorical patterns they illustrate apply across the discipline.

An Engineering Corpus: Why It Matters

Most corpus-based writing guides draw on social science or biomedical research articles. These guides are valuable, but engineering writing differs in several important ways.

Structure. Social science articles almost universally follow the IMRDC template: Introduction, Methods, Results, Discussion, Conclusion. Engineering articles do not. In the corpus, only 1.2% of papers use a full IMRDC structure. The vast majority organise their body under technical headings, that is, headings that name systems, methods, models, or analytical stages rather than the stages of research. A typical engineering article might be organised as Introduction → System Model → Numerical Simulation → Parametric Study → Conclusions, with no section labelled Methods, Results, or Discussion. This means that much of the advice in IMRDC-based writing guides: put your methods in the Methods section, save your interpretation for the Discussion etc does not directly apply.

The develop→test→validate arc. Where social science papers are organised around an empirical cycle (hypothesis → method → data → interpretation → conclusion), engineering papers are organised around a technical contribution: develop a method, model, or system; test it computationally or experimentally; validate it against benchmarks or established data; and draw conclusions about its performance. This arc shapes everything: which sections the paper has, what tense predominates, where figures cluster, and how certainty is managed.

Visual and mathematical density. Engineering papers communicate through figures, tables, and equations to a far greater degree than papers in many other disciplines. The corpus contains 2,751 figure references, 550 table references, and 149 equation references, an average of roughly one figure reference every 220 words across the corpus, and considerably more in the technical body sections. A substantial proportion of the paper's rhetorical work is done not through text argument but through the narration of visual and mathematical displays: introducing a figure, directing the reader's attention to a feature, and stating what the figure means.

Validation as a rhetorical move. Engineering papers make a claim (the proposed method works) and then provide evidence (it matches experimental data, it outperforms alternatives, it agrees with benchmarks). This validation move, present in 55% of corpus articles, though only 11% use an explicit Validation heading, has no clear parallel in social science or humanities writing. It is closer to a proof-of-concept demonstration than to the interpretive Discussion of other disciplines.

We talk about moves or rhetorical moves. These are just the functional elements that build up to tell a story in the engineer research article. For instance, sometimes in an article it is necessary to provide some validation of a model or similar. We simply label that part of the article as a validation move. And we see that around half the articles have that move.

The "proposed" frame. Engineering writers typically frame their contribution as a method, model, system, or design rather than as an argument or interpretation. The phrase *the proposed method* alone appears 73 times in the corpus; *a novel* -- appears 113 times; *a new* -- appears 152 times. The engineering niche is a thing built, not a position argued.

This book addresses these differences directly. Each unit analyses the rhetorical patterns of engineering papers as they actually appear. These patterns are described as they appear (under technical headings, around figures and equations, through the develop→test→validate arc etc.) rather than forcing them into a framework designed for other disciplines.

Why Corpus Evidence Helps Engineers Write

Most academic writing textbooks work from the top down: they present a model of what good writing looks like and ask students to reproduce it. Corpus analysis works from the bottom up: it examines what published writers actually produce and identifies the patterns that emerge.

This matters for you as an engineering writer for three reasons.

First, corpus evidence replaces intuition with frequency. Experienced writers often cannot articulate why they make particular choices; they "just know" that an Introduction needs to establish a gap, or that a Conclusion should restate the contribution before pointing to future work. Corpus data makes this tacit knowledge explicit. When we show that *however* appears roughly once every 700 words in Introductions, we are quantifying a pattern that experienced writers sense but cannot state. When we show that *this paper* appears at 16.9/10k in Conclusions but only 8.7/10k in Introductions, we are making visible a shift in rhetorical function that experienced readers feel but rarely articulate.

Second, corpus evidence reveals the architecture of the paper. Individual engineering articles look different from one another: different systems, different methods, different applications. But at the level of rhetorical structure, they are remarkably similar. Introductions show "gaps." Technical body sections present and test solutions. Conclusions claim contributions and point to the future. Figure references peak in the body and vanish in the Conclusion. Boosting language peaks in the Conclusion at twice the Introduction rate (29.8/10k vs. 14.7/10k). The corpus makes this shared architecture visible.

We will go through these structures systematically in this book.

Third, corpus evidence shows range, not just norms. For any given rhetorical function, the corpus reveals multiple ways of accomplishing it. A research gap can be signalled through technical limitation (*the main drawback of this approach...*), practical need (*there is an urgent need for...*), absence of research (*has not been fully explored...*), or performance challenge (*remains a significant challenge...*). Seeing this range helps you choose the strategy that best fits your study.

How to Use This Book: Conceptualise First, Then Find the Language

A corpus-based writing book carries a risk: you may treat it as a phrase bank, copying constructions wholesale without understanding the rhetorical work those constructions perform. This would be a mistake. The examples in this book are not templates to be filled in; they are evidence of how published writers have solved particular rhetorical problems.

Before consulting any section of this book, you should have already done the conceptual work:

Before writing your Introduction, you should be able to answer: *What is the specific technical gap or problem my study addresses? Why does it matter? What exactly does my study do about it?* Only then should you look at how other writers have signalled gaps and announced their contributions, not to copy their words, but to find constructions that fit your conceptualisation.

Before writing your theoretical or modelling sections, you should be able to answer: *What are my governing equations? What assumptions am I making, and why? What is the logical sequence of my derivation?* The corpus shows you how other writers present equations, state assumptions, and narrate derivations; but the mathematical logic of your own model must come first.

Before writing your experimental or computational sections, you should be able to answer: *What did I do? What apparatus, software, or procedures did I use? What parameters did I specify, and why?* The corpus shows you how other writers describe methods and justify choices; but your own methodological decisions must already be clear.

Before writing your results and validation sections, you should be able to answer: *What are my main findings? What figures or tables best display them? Against what benchmarks am I validating, and what counts as good agreement?* The corpus shows you how other writers narrate figures, report quantitative outcomes, and characterise agreement; but your interpretation of your own results must be original.

Before writing your Conclusion, you should be able to answer: *What did I contribute? What are the practical implications? What comes next?* The corpus shows you how other writers summarise, claim significance, and identify future work; but the assessment of your own study's contribution must be yours.

The process is simple: **conceptualise → consult → compose.** Decide what you want to say, look at how others have said similar things, then write in your own voice using structures that serve your argument. The corpus gives you a repertoire of options. Your technical contribution determines which options to choose.

The analysis in this book reflects general trends across engineering. Individual papers and sub-disciplines will vary, so don't blindly follow the structures outlined here. Keep the conceptualisation of your particular study at the forefront. That is what determines the appropriate structure.

What the Corpus Reveals: A First Look at Section Proportions

Before examining specific structures, it is worth seeing the paper as a whole. The corpus reveals that the major functional components of the engineering paper are not equal in length:

Section type	Share of paper	Words in a 3,800-word paper	Character
Introduction	34.4%	~1,310	Longest section; functions as literature review
Technical body sections	40.2%	~1,530	Core technical content
Conclusion	7.5%	~285	Short and formulaic
Other (Methods, Results, Discussion, Background)	17.9%	~680	Variable by paper type

Three observations stand out.

First, the Introduction is long: over a third of the paper. Engineering Introductions do substantial work: they establish the field, review the literature (usually without a separate Literature Review section, only 11% of corpus articles have one), identify the gap, and announce the study's contribution. If your Introduction is a single paragraph, you are drastically underserving this section.

Second, the Conclusion is short. At approximately 315 words on average (7.5% of the paper), the engineering Conclusion is the most compressed functional section. It restates what was done, summarises key results with numerical specifics, and often points to future work typically in two to four paragraphs. This compression is a feature, not a shortcoming: by the time the reader reaches the Conclusion, the evidence has already been presented and validated.

Third, the technical body sections carry 40% of the paper. This is where the engineering paper's real work happens: the modelling, the testing, the figure-by-figure presentation of results, the comparison with benchmarks. These sections are typically organised under technical headings (*Finite Element Analysis*, *Computational Domain and Boundary Conditions*, *Experimental Programme*) rather than functional labels.

Five Paper Types

A distinctive feature of this book is that it addresses the full range of engineering paper structures. The corpus reveals five macrostructure families, each reflecting a different type of technical investigation.

Path B (Theory/Simulation) is the most common at 28.2%. These papers develop a mathematical model or computational method, test it through simulation, and draw conclusions from the simulated results — all without physical experimentation. Path C (Experimental) is the closest to the traditional IMRDC template, and the most likely to carry functional headings like Methods and Results. Path A (Theory + Experiment) combines both, typically presenting a theoretical development followed by experimental validation.

Path	Structure	Prevalence	Typical fields
A: Theory + Experiment	Intro → Model/Theory → Experimental → [Validation] → Conclusion	21.5%	Aerospace, Mechanical, Materials
B: Theory/Simulation	Intro → Model/Theory → Simulation → [Parametric Study] → Conclusion	28.2%	Control, Computational, Electrical
C: Experimental	Intro → [Methods] → Experimental → Results [+ Discussion] → Conclusion	25.2%	Civil, Manufacturing, Chemical
D: Analysis/Evaluation	Intro → [Background] → Analysis/Evaluation → [Discussion] → Conclusion	11.0%	Cross-disciplinary, Assessment
E: Thematic	Intro → Domain-specific body sections → Conclusion	14.1%	Multidisciplinary, Conceptual

Throughout this book, every unit presents data relevant to all five paths. The rhetorical moves are the same across paper types: gap-signalling, assumption-stating, figure-narrating, contribution-claiming, but they are distributed differently depending on the paper's structure. If you are writing a Path B simulation paper, for example, you will not need a Methods section in the traditional sense, but you will need to describe your computational setup, state your assumptions, present your governing equations, and validate your results.

How This Book Is Organised

The units in this book are presented in four groups, moving from the macrostructure of the paper to the internal architecture of individual sections to features that arc across the whole paper to a final synthesis.

Unit 1: Paper Macrostructure — The five structural paths, heading design, and how to navigate this guide based on your paper type.

Unit 2: Writing the Abstract — Five-move structure, context opening, novelty claims, validation preview.

Part 1: Rhetorical Moves Within Sections: What writers do within each section.

These units describe the characteristic rhetorical "moves", the recurring steps that writers take within each type of section. They show, for example, that Introduction writers signal technical deficiency before proposing their solution, that modelling sections thread assumptions throughout the mathematical development, and that Conclusion writers claim contribution before identifying future work. Each move is illustrated with corpus frequencies and example sentences.

Unit 3: Introduction Moves — Territory, gap, and niche in engineering Introductions.

Unit 4: Presenting Mathematical and Theoretical Development — Assumptions, equations, models, derivations.

Unit 5: Describing Experimental and Computational Methods — Apparatus, software, procedures, parameters.

Unit 6: Reporting and Interpreting Results — Figure narration, quantitative reporting, commentary.

Unit 7: The Validation Move — Comparison, agreement, discrepancy: how to show that it works.

Unit 8: Discussion Moves — Interpretation, comparison with literature, limitations.

Unit 9: Conclusion Moves — Summary, contribution, future work.

Part 2: Opening and Closing Strategies: How writers begin and end sections.

Unit 10: First and Last Sentences of Each Section

Part 3: Cross-Section Arcs: Features that change systematically across the paper.

These units examine features that shift from section to section, creating trajectories that run through the whole paper. Figure references peak in the body and vanish in the Conclusion. Tense shifts from present (established knowledge) to past (what was done) and back to present (what it means). Boosting language doubles from Introduction to Conclusion. Together, these arcs reveal the paper's underlying argumentative logic.

Unit 11: Tense Distribution Across Sections

Unit 12: Narrating Figures, Tables, and Equations

Unit 13: Citation Distribution and Integration

Unit 14: Hedging and Boosting

Unit 15: Authorial Presence and Self-Reference

Capstone: The Paper as a Single Technical Argument

The final unit synthesises all previous analyses to show how the entire paper serves a single macro-argument: there is a technical problem, and we solve it. It draws together the move structures, the tense arc, the visual apparatus, and the hedging arc to show how every linguistic choice contributes to this overarching case.

Unit 16: The Paper as Argument

Reading the Data in This Book

Throughout this book, linguistic features are reported as frequencies per 10,000 words. This normalised rate allows direct comparison across sections of different lengths. A frequency of 15.0/10k means the feature appears roughly 1.5 times in every thousand words, or once every 670 words.

The tables in each unit present frequencies broken down by section type or by position in the paper. The key column to watch is usually the one with the highest rate: which section or position shows the peak? This peak tells you where the feature does its most concentrated work.

Rate per 10,000 words	Rough equivalent	Example from this corpus
1.0/10k	Once every 1,000 words (~once per page)	suffer from in Introductions (0.9/10k)
5.0/10k	Once every 200 words (~once per long paragraph)	this work in Conclusions (5.5/10k)
10.0/10k	Once every 100 words (~once per short paragraph)	however in Conclusions (12.3/10k)
15.0/10k	Once every 67 words (~every three sentences)	however in Introductions (13.8/10k)
30.0/10k	Once every 33 words (~every sentence or two)	Figure refs in Introductions (29.7/10k)
60.0/10k	Once every 17 words (~once per sentence)	Figure refs in body sections (60.0/10k)

Some practical benchmarks for interpreting frequencies in engineering writing:

When the text says a feature "peaks" in a particular section, this means that section has the highest normalised frequency, not necessarily the highest raw count. Because sections differ in length, a shorter section with a high rate may have fewer total instances than a longer section with a lower rate.

A Note on Prescription and Description

The corpus shows you what is common, not what is compulsory. Your task is not to reproduce these patterns mechanically but to understand the rhetorical problems they solve, and then to solve those problems in your own way, for your own study. What we are doing is making the flow of Engineering articles explicit. You will have some idea of that flow from your reading, but your sense of that structure may be a little vague. Making these patterns explicit will help you become a more proficient in writing according to the practices of your discipline.

The patterns described here represent the collective practice of over 160 published engineering researchers. They are benchmarks, not rules. If your Conclusion is longer than the corpus average of 315 words, it may be that your study has more complex implications to draw out, or that your sub-discipline expects a more discursive ending. If your Introduction has fewer citations than the corpus norm, it may be that your topic is sufficiently novel that there is less prior work to review. The numbers here give you a reference point; your judgement determines how closely to follow the patterns presented here.

What the corpus can do is alert you when your writing diverges from disciplinary norms. For instance, if your Introduction has no gap-signalling language, the corpus data suggests you may be working against reader expectations. This is where the numbers are most useful, not as targets to hit, but as signals that something may need attention.

Unit 1: Research Paper Macrostructure and Your Path Through the Units

Engineering Corpus — 163 articles analysed

The Key Finding: Engineering Papers Do Not Follow IMRDC

Most guides to academic writing assume that research papers follow the Introduction–Methods–Results–Discussion–Conclusion (IMRDC) structure. The IMRDC structure is found in many disciplines, but not so much in engineering.

Feature	Percentage of 163 articles
Has an Introduction heading	93.9%
Has a Conclusion/Conclusions heading	88.3%
Has Modelling/Theory/Simulation content (heading or sub-heading)	49.7%
Has Methods or Methodology heading	33.1%
Has Experimental content (heading or sub-heading)	20.9%
Has a Results heading	17.8%
Has a Discussion heading	16.6%
Has combined Results and Discussion	13.5%
Full IMRDC (M + R + D all present)	1.2%
Has domain-specific technical headings	85.9%

Our analysis of 163 engineering research articles reveals that **only 1.2% use a full IMRDC structure**. The remaining 99% organise at least part of their body using **technical headings**: topic-driven labels that reflect the study's methodology, system, or analytical logic rather than the generic stages of a research process.

The universal or near-universal elements are the the Abstract (100%), the Introduction (93.9%) and the Conclusion (88.3%). Everything in between is variable. Nearly half the corpus includes modelling or theoretical development. A third has some form of explicit Methods section. But the conventional IMRDC sequence, where these appear as separate, labelled, ordered sections, is almost never used.

This guide teaches the rhetorical moves that appear in all engineering papers: establishing a territory, identifying a gap, presenting a model, describing an experiment, reporting results, validating against benchmarks, and claiming a contribution. (We illustrate what these moves mean in the following chapters.) These moves are universal. What differs across engineering is how they are organised; It could be under functional headings (Methods, Results, Discussion) or under technical headings that reflect the study's own logic.

Five Paths Through the Engineering Paper

Path	Structure	Preval ence	Typical fields
A: Theory + Experiment	Intro → Model/Theory → Experimental → [Validation] → Conclusion	22.7%	Aerospace, Civil/Structural, Mechanical
B: Theory/Si mulation	Intro → Model/Theory → Simulation → [Parametric Study] → Conclusion	25.2%	Aerospace, General Engineering, Control/Systems
C: Experimental	Intro → [Methods] → Experimental → Results [+ Discussion] → Conclusion	23.9%	Civil/Structural, Indus trial/Manufacturing, Aerospace
D: Analysis/ Evaluation	Intro → [Background] → Analysis/Evaluation → [Discussion] → Conclusion	8.6%	Cross-disciplinary
E: Thematic	Intro → Domain-specific body sections → Conclusion	19.6%	Aerospace, Control/Systems, General Engineering

The corpus reveals five dominant macrostructure families. Your research question, target journal, and disciplinary tradition will determine which path you follow.

The remaining articles are variations on these patterns.

PATH A: THEORY + EXPERIMENT (22.7%)

These papers develop a theoretical or computational model **and** validate it experimentally. The body typically moves from mathematical formulation to physical testing and the full develop→test→validate arc within a single paper. This is the most structurally complex path, and the most likely to include explicit Validation sub-sections within it.

Heading	Type	Typical content
Introduction	Functional	Territory, gap, purpose (Unit 4)
[Theory/Model section]	Technical	Mathematical formulation, assumptions (Unit 5)
[Experimental section]	Technical	Apparatus, specimens, procedures (Unit 6)
[Results / Validation]	Technical or Functional	Comparison of model with experiment (Units 6–6)
Conclusions	Functional	Summary, contribution, future work (Unit 10)

What this looks like:

Corpus examples of heading sequences:

"A Statistical Energy Analysis (SEA) model of a fuselage section..."

Introduction → The Basics of SEA Theory → SEA Model of a Fuselage Section → Results → Conclusion

"A novel cryogenic machining concept based on a lubricated liquid carbon dioxide..."

Introduction → Concept → Theoretical Background → Experimental Proof of Concept → Conclusions and Future Work

"Characterisation of turbine behaviour for an engine overspeed prediction..."

Introduction → Methodology → Computational Domain and Boundary Conditions → Automation Framework and Flow Solution → Results and Discussion → Conclusions

How to use this guide for Path A papers: Read Units 3 (Introduction) and 9 (Conclusion) for your opening and closing. Unit 4 (Mathematical/Theoretical Development) covers your modelling sections. Unit 5 (Experimental/Computational Methods) covers your procedural sections. Unit 7 (The Validation Move) is particularly important for Path A, since the comparison between theory and experiment is the paper's central theme. Units 6 and 8 (Results, Discussion) apply wherever you report and interpret findings.

Path B: Theory/Simulation (25.2%)

This is the most common path. These papers develop a mathematical model, computational method, or algorithm and test it through simulation without any physical experimentation. The body moves from formulation to computational testing to performance evaluation. Validation, if present, is against published data, analytical solutions, or benchmark cases rather than the authors' own experiments.

Heading	Type	Typical content
Introduction	Functional	Territory, gap, purpose (Unit 4)
[Model/System section]	Technical	Mathematical formulation, governing equations (Unit 5)
[Simulation/Numerical section]	Technical	Computational setup, parameters (Unit 6)
[Results / Evaluation]	Technical or Functional	Performance, parametric analysis (Unit 7)
Conclusions	Functional	Summary, contribution, future work (Unit 10)

What this looks like:

Corpus examples:

"A method to improve an electric vehicle's range: Efficient Cruise Control..."

Introduction → System Model → Efficient Cruise Control Design → Adaptive Cruise Control Design → Evaluation of Efficient Cruise Control → Conclusions

"An improved multiple model particle filtering approach..."

Introduction → GMTI Tracking Models → Numerical Simulations → Conclusions

"Aeroelastic simulations of stores in weapon bays..."

Introduction → Model Geometry and Mesh Generation → Results and Discussion → Conclusions

How to use this guide for Path B papers: Read Units 3 and 9 for Introduction and Conclusion. Unit 4 is essential; it is the theoretical development is the heart of your paper. Unit 5 covers computational setup. Unit 6 covers results reporting. If you validate against published data, Unit 7 applies.

Path C: Experimental (23.9%)
These papers report primarily experimental work.

They are the closest to the traditional IMRDC template and the most likely to carry functional headings like Methods, Results, and Discussion, though many still use technical headings for the procedural sections.

Heading	Type	Typical content
Introduction	Functional	Territory, gap, purpose (Unit 4)
[Methods/Experimental section]	Functional or Technical	Materials, apparatus, procedures (Unit 6)
[Results section]	Functional or Technical	Findings, figure narration (Unit 7)
[Discussion] (optional)	Functional	Interpretation, comparison, limitations (Unit 9)
Conclusions	Functional	Summary, contribution, future work (Unit 10)

What this looks like:

Corpus examples:

"Areal surface texture data extraction from X-ray computed tomography..."

Introduction → Methodology → Dimensional Artefact Results → Discussion → Conclusions → Future Work

"Drill-exit temperature characteristics in drilling of UD and MD CFRP..."

Introduction → Results and Discussion → Conclusions

"A novel process-linked assembly failure model for adhesively bonded..."

Introduction → Nonlinear Correlation Analysis → Experiments → Conclusions

How to use this guide for Path C papers: Follow Units 3, 4, 5, 8, and 9 in sequence; these map most directly to your sections. Unit 7 (Validation) applies if you compare results against established models or standards.

Path D: Analysis/Evaluation (8.6%)

These papers present an evaluation, assessment, or comparative analysis rather than developing a new method or reporting original experiments. They typically apply established techniques to a specific problem and interpret the outcomes.

Corpus examples:

"Application of Monte Carlo simulation for estimation of uncertainty..."

Introduction → Results and Discussion → Conclusions

"Chatter avoidance via structural modification of tool-holder geometry..."

Introduction → Background Theory → Discussion → Conclusion

"Deterioration of cracks in onshore wind turbine foundations..."

Introduction → Results → Discussion → Conclusion

How to use this guide for Path D papers: Read Units 3 and 9 for Introduction and Conclusion. Unit 8 (Discussion) is particularly important, as the interpretive work is the paper's primary contribution. Units 6 (Results) and 12 (Narrating Figures) apply to your analytical sections.

Path E: Thematic (19.6%)

These papers organise their body entirely under domain-specific headings that reflect the study's conceptual or investigative structure. There are no Methods, Results, or Discussion labels. Instead, the functions of those sections are embedded within the thematic sections, much as in other disciplines, though the content is technical.

Corpus examples:

"A comprehensive assessment of collision likelihood in Geosynchronous Earth Orbit..."

Introduction → The Current Public Space Population → Characterization of GEO Close Approaches → GEO Collision Likelihood and Encounter Rate Estimation Techniques → External Research to Date → Overall GEO Active Satellite Collision Likelihood → Consequence of GEO Collision → Conclusions

"An exploration of the effectiveness of artificial mini-magnetospheres..."

Introduction → The Hazard → Estimating the Requirements of the On Board Hardware → Evidence for the Processes from Other Fields → Summary → Conclusions

"An extended linear quadratic regulator for LTI systems..."

Introduction → Problem Statement → Conclusions

How to use this guide for Path E papers: Read Units 3 and 9 for Introduction and Conclusion. For your thematic body sections, draw on Units 4-7 as needed; the moves described in those units will appear within your thematic sections, not as separate sections. Unit 12 (Narrating Figures) applies directly regardless of path.

Designing Technical Headings

The corpus contains 1,326 technical headings (main and sub-headings, excluding Introduction and Conclusion). Analysis reveals four dominant naming patterns:

Principles for effective technical headings:

1. Headings should name the technical content, not just the topic. Compare: *"Model"* (too vague) vs. *"Finite Element Model of the Fuselage Section"* (specific system + specific method). The heading should tell the reader what kind of work the section performs.

2. Headings should form a logical sequence. Read your headings in order. Do they trace the develop→test→validate arc? The sequence should have a progression that mirrors the paper's logic: typically from formulation to implementation to evaluation.

3. Headings should be parallel in form. If your first body heading is a noun phrase, maintain that form throughout. Mixing gerunds, questions, and noun phrases within a single paper creates a disjointed reading experience.

4. Technical sections still contain IMRDC moves. A section titled *"Computational Domain and Boundary Conditions"* will contain Methods moves (specifying procedures, justifying choices). A section titled *"Numerical Simulations"* will contain Results moves (reporting outcomes, narrating figures). The moves are present; only the labels differ. This is why the units in this book are organised by rhetorical move rather than by section label.

Strategy	Examples	Frequency
Noun phrase (the dominant pattern)	*Fuselage section configurations*; *Mesh convergence*; *Close conjunction statistics*	~69%
X of Y / X for Y	*Optimization of the power block*; *Sizing of the solar block*; *Decision support tool for waste heat recovery*	~25%
Process / gerund	*Modelling of adhesive flow*; *Calculating stand-off distance*; *Estimating atmospheric density*	~5%
Numbered / labelled	*Method 1: statistical evaluation of JSpOC CDMs*; *Step 1: waste heat survey*; *Configuration A*	~2%

How Many Body Sections?

The typical engineering paper has 2 main body sections between the Introduction and Conclusion. However, this counts only main headings. The mean number of sub-headings is 6.2 per article, and 91.4% of articles use sub-headings. The real organisational depth of the engineering paper is in the sub-sections.

Body section count varies by path: Path A (Theory + Experiment) averages 2.2 main body sections, Path B (Theory/Simulation) averages 1.9, Path C (Experimental) averages 2.2, while Path E (Thematic) averages 1.4 main sections but often uses more sub-headings to create internal structure.

Statistic	Value
Mean main body sections (excl. Intro and Conclusion)	2.0
Median	2
Most common	2 (26.4% of articles)
Range	0–8
Articles with 0 body sections (Intro + Conclusion only)	19.0%
Articles with 1–2 body sections	50.9%
Articles with 3–4 body sections	24.5%
Articles with 5+ body sections	5.5%

Identifying Your Path

Before you begin writing, examine 3-5 recent articles in your target journal. For each, note:

1. Are the body section headings functional (Methods, Results, Discussion) or technical (system-specific, method-specific)?

2. Does the paper develop a mathematical model or theoretical framework? If so, is it validated experimentally, computationally, or against published data?

3. Is there a separate experimental/methods section, or are procedures described within technical sections?

4. Is there a standalone Discussion section, or does interpretation occur throughout the body?

5. Does the Conclusion include a Future Work component?

Your answers will place you on one of the five paths:

Technical headings, model + experiment = Path A

Technical headings, model/simulation, no physical experiment = Path B

Functional or technical headings, primarily experimental = Path C

Functional headings, analysis/evaluation focus = Path D

Domain-specific headings throughout, no clear M/R/D = Path E

Guide Map: Which Units to Read

The table below may be a little hard to read. It is an attempt to relate the units in the book to the type of paper you are aiming to write. You choose one of the paths and see how to navigate through the book. Ess stands for Essential; Emb stands for Embedded.

A note on "embedded": When a unit is marked as "embedded," this means the rhetorical moves described in that unit still appear in your paper; they are simply distributed across your technical body sections rather than concentrated in a single functional section. Read the unit to learn the moves, then deploy them where they naturally fit within your technical structure.

Unit	Topic	Path A	Path B	Path C	Path D	Path E
2	Writing the Abstract	Ess	Ess	Ess	Ess	Ess
3	Introduction Moves	Ess	Ess	Ess	Ess	Ess
4	Mathematical/Theoretical Development	Ess	Ess	If applicable	Background	Emb
5	Experimental/Computational Methods	Ess	Computational setup	Ess	If applicable	Emb
6	Reporting and Interpreting Results	Ess	Ess	Ess	Ess	Emb
7	The Validation Move	Ess	If validating	If comparing	Occasional	Emb
8	Discussion Moves	If separate	Emb	Ess	Ess	Emb
9	Conclusion Moves	Ess	Ess	Ess	Ess	Ess
10	First and Last Sentences	Ess	Ess	Ess	Ess	Ess
11	Tense Distribution	Ess	Ess	Ess	Ess	Ess
12	Narrating Figures, Tables, Equations	Ess	Ess	Ess	Ess	Ess
13	Citation Distribution	Ess	Ess	Ess	Ess	Ess
14	Hedging and Boosting	Ess	Ess	Ess	Ess	Ess
15	Authorial Presence	Ess	Ess	Ess	Ess	Ess
16	The Paper as Argument	Ess	Ess	Ess	Ess	Ess

Unit 2: Writing the Abstract

Engineering Corpus — 155 abstracts analysed, 30,225 words

Overview

The Abstract is the most-read part of your paper, and the part that determines whether readers continue to look at the full text. Our analysis reveals a consistent five-move structure, a median length of **188 words** (~8 sentences), and a strong preference for opening with context (70%) over purpose (28%).

Abstract Length

Key finding: Engineering abstracts cluster between 150 and 250 words. The average contains approximately 8 sentences.

The Five-Move Structure

Engineering abstracts follow a compressed version of the paper itself:

Move	Function	Share of abstract
1. Context	Why the topic matters	~1–2 sentences
2. Gap/Problem	What is missing or needed	~1 sentence
3. Purpose	What this paper does	~1–2 sentences
4. Method	How it was done (briefly)	~1–2 sentences
5. Results/Contribution	What was found or achieved	~2–3 sentences

Move 1: Context (Opening)

Key finding: 70% of abstracts open with context: a broad statement about why the topic matters. Only 28% open with *"This paper proposes…"*

Corpus examples:

> *"Concentrated Solar Power (CSP) technologies are now fast advancing and competitive."* (context: technology landscape)
> *"Comfort plays an increasingly important role in the interior design of airplanes."* (context: user need)
> *"One third of energy consumption is attributable to the industrial sector, with as much as half ultimately wasted as heat."* (context: problem scale)

Move 2: Gap/Problem

However/but appears at 28.2/10k in abstracts, which is substantially higher than in the full corpus. The gap is compressed to a single sentence.

Corpus examples:

> *"However, knowledge behind the interaction between curing process parameters and the failure of polymer composite bonded joints is not keeping pace with the market."* (however - knowledge gap)
> *"Yet for all of its importance, a comprehensive assessment of GEO collision likelihood is difficult to do and has never been done."* (yet - unresolved problem)

Writer's note: The Abstract gap is brief, just one sentence. Use *however*, *but*, or *yet* to pivot from context to the gap.

Move 3: Purpose

Construction	Per 10k	Function
novel / new / for the first time	28.9	Novelty claim
is proposed / is presented / is developed	12.7	Passive method announcement
we propose / present / develop	7.8	Active method announcement
this paper/study/work [verb]	7.5	Paper as subject
in this paper/study/work	7.2	Framing

Key finding: Novelty language (*novel/new* at 28.9/10k) is even denser in abstracts than in Conclusions (20.4/10k). The Abstract is the place to claim novelty.

Corpus examples:

> *"This paper addresses the problem of optimal control for semi-active vehicle suspensions." (this paper addresses)*
> *"A decision support tool is presented which computes the compatibility of waste heat source(s) and sink(s)." (is presented)*
> *"A novel nonlinear correlation analysis has been employed and applied to experimental data." (novel* - novelty + method)
> *"In this paper, we employ six independent and diverse assessment methods to estimate GEO collision likelihood." (in this paper, we employ)*

Move 4: Method (Brief)

The method description is compressed: naming the approach without procedural detail.

Corpus examples:

> *"A time-domain optimal control approach is adopted in which ride and handling are modelled as exogenous disturbances acting on the vehicle." (approach named + key feature)*
> *"A control algorithm is derived from a solution of the stochastic Hamilton-Jacobi-Bellman equation for the finite horizon case." (algorithm derivation)*

Writer's note: Name the approach (1-2 sentences), don't describe the procedure. Not *"The specimen was mounted in a tensile testing machine and loaded at 2 mm/min"* but *"Tensile testing was performed on adhesively bonded composite specimens."* You will provide the details later.

Move 5: Results/Contribution

Key finding: Nearly half of abstracts close with a specific result claim. Validation language (*good agreement, validated* at 5.8/10k) confirms that readers expect to know in the Abstract whether the method was validated.

Pattern	Frequency	Function
Result / finding claim	48%	Specific finding with numbers
Other (method/context)	35%	
Future work	9%	Projecting next steps
Contribution / novelty	7%	Asserting the contribution

Corpus examples:

> *"The advantages of the approach are demonstrated experimentally on a test vehicle performing a steering manoeuvre on a bumpy roundabout." (demonstrated experimentally)*
> *"Substantial improvement in plant energy efficiency together with reduction in the payback time for heat recovery has been demonstrated." (improvement demonstrated)*
> *"These results are within a factor of 3.5 of each other."* (specific numerical finding)

A Complete Abstract Template

Context (1-2 sentences): *"[Topic] is [important/challenging/growing]. [Broader significance]."*

Gap (1 sentence): *"However, [what is missing/unsolved/limited]."*

Purpose (1-2 sentences): *"This paper [proposes/presents/develops] a [novel/new] [method/model/approach] for [purpose]."*

Method (1-2 sentences): *"[Approach] is [adopted/employed]. [Key feature of the method]."*

Results (2-3 sentences): *"Results [show/demonstrate] that [specific finding with numbers]. [Validation outcome]. [Key implication]."*

Key Frequencies: Abstract vs. Full Paper

Key finding: The Abstract is a compressed, high-density version of the paper. Every sentence carries argumentative weight.

Construction	Abstract	Full paper	Ratio
novel / new	28.9/10k	~5/10k	6× denser
however / but	28.2/10k	~15/10k	2× denser
this paper [verb]	7.5/10k	~2.5/10k	3× denser
results show/demonstrate	6.3/10k	~3.8/10k	1.7× denser
validated / good agreement	5.8/10k	~1.5/10k	4× denser

Implications for Writing

Open with Context, Not Purpose

70% of abstracts open with a broad contextual statement. Give the reader one sentence of context before announcing your study.

Claim Novelty Explicitly

At 28.9/10k, novelty language is at its densest in the Abstract. Use *novel*, *new*, or *for the first time* if warranted.

Include Specific Numbers in Results

Nearly half of abstracts close with a specific result claim including numbers. *"Energy consumption was reduced by 23.56%"* is more compelling than *"Energy consumption was reduced."*

Preview Your Validation

Validation language at 5.8/10k confirms that readers expect to know whether the method was validated. Include one sentence on validation outcome.

Keep to 150-250 Words

The median is 188 words. Exceeding 300 suggests too much detail; below 100 is likely to be too little.

Unit 3: Rhetorical Moves in the Introduction

Engineering Corpus — Introduction Sections (151 Introductions, 209,705 words)

Overview

Swales' (1990, 2004) Create a Research Space (CARS) model describes three rhetorical moves in research article Introductions: establishing a territory, establishing a niche, and occupying the niche. The corpus data from engineering research articles illustrates these moves, but with a distribution and character that differ markedly from the humanities.

The engineering Introduction accounts for **34.4%** of the typical paper; it is the largest single section. In a 3,800-word paper, this translates to roughly 1,310 words. The Introduction typically functions as the only place where prior work is reviewed at length: only 11% of corpus articles have a separate Literature Review or Background section.

Before You Write: Conceptualising Your Introduction

Before looking at the move patterns below, take ten minutes to think through the argumentative logic of your Introduction.

Step 1: Define Your Territory

Write a short paragraph (3-4 sentences) answering: What broad technical field does your study belong to? Why does this field matter; what practical or theoretical problem does it address? What have other researchers already established?

Do not worry about phrasing yet. Write in your own words, as if explaining to a colleague over coffee.

Step 2: Identify Your Niche

Write 2-3 sentences answering: What is specifically missing, inadequate, or problematic in the current state of the art? Is this a *technical limitation* (current methods cannot achieve adequate performance), a *practical need* (industry demands something that does not yet exist), a *research absence* (nobody has studied this), or an *extension opportunity* (existing work can be improved or generalised)?

This step is the most important. If you cannot state your niche clearly in plain language, your Introduction will lack focus.

Step 3: State Your Response

Write 1-2 sentences answering: What exactly does your study do about the problem you identified? What is the specific method, model, system, or design you propose?

Step 4: Check the Connection

Read your three paragraphs in sequence. Does the niche follow logically from the territory? Does your response directly address the niche? If there are gaps in the logic, revise before moving on.

Move 1: Establishing a Territory

This move claims that the research area is important, active, and well-established. In engineering, territory-establishing tends to open with a broad significance statement about the technical domain, then narrow quickly to the specific problem.

Key finding: Move 1 density in engineering (7.4/10k) is low — comparable to the humanities (8.5/10k). Engineering writers establish territory quickly, often in one or two sentences, then move rapidly to the problem.

Usage	Count	Per 10k	Function
[important/critical/key] + [role/issue/challenge]	50	2.4	Claiming topic importance
[well/widely] + [established/documented/studied]	25	1.2	Citing established knowledge
X has become/emerged/gained/attracted	22	1.0	Showing growing significance
in recent years/decades	17	0.8	Temporal framing
plays a/an [important/critical] role	16	0.8	Role-claiming
[extensive/growing] + [literature/research]	16	0.8	Showing scholarly activity
[growing/increasing] + [interest/attention]	9	0.4	Showing research momentum
Total Move 1	155	7.4	

Corpus examples:

> *"Planar nanopositioning platforms play a crucial role in precise and accurate nanoscale positioning." (topic importance)*
> *"With its ability to handle constraints and nonlinear dynamics systematically, NMPC has become widely used in process and chemical applications." (growing significance)*
> *"In recent years, the study of mitigation solutions regarding tackling the rising seas challenges has dramatically increased." (temporal framing)*

Move 2: Establishing a Niche (the Gap)

This move identifies what is missing, problematic, or inadequately addressed. In engineering, this is overwhelmingly the longest and most heavily marked move.

Key finding: Move 2 density in engineering (64.1/10k) is dramatically higher than in the humanities (24.7/10k). Engineering writers invest enormous effort in establishing the niche. The top two markers tell the story: *need for / demand for / require* (14.9/10k) and *however* (13.8/10k). Together, these account for nearly half of all niche-signalling.

Usage	Count	Per 10k	Function
need for / demand for / require	313	14.9	Stating practical necessity
however	289	13.8	Pivoting to the gap
limited/limitation	164	7.8	Signalling deficiency
complex/complicated/computationally expensive	126	6.0	Flagging technical difficulty
challenge/challenging	87	4.1	Naming the problem
although / even though	83	4.0	Conceding before pivoting
cannot / unable to / fail to	60	2.9	Stating inability
gap	48	2.3	Naming the gap explicitly
despite	32	1.5	Conceding before pivoting
lack/absence/paucity of	31	1.5	Signalling absence
has/have not been [studied/investigated]	23	1.1	Noting research absence
little/few/no + research/studies	20	1.0	Stating the gap directly
suffer from	18	0.9	Technical deficiency
drawback/disadvantage/shortcoming	17	0.8	Naming specific problems
overlooked/neglected/underexplored	17	0.8	Noting scholarly neglect
Total Move 2	~1,345	64.1	

The critical difference from the humanities is in *what kind of gap* is signalled. Humanities writers favour research absence (*little research has examined…*). Engineering writers favour **technical limitation** (*suffer from, drawback, cannot*) and **practical need** (*need for, is essential*). The gap in engineering is typically not "nobody has studied this" but "existing approaches do not work well enough."

Corpus examples:

> *"One way to overcome this drawback is to use adaptive/approximate dynamic programming (ADP) to obtain an approximate solution of the HJB equation." (technical limitation)*
> *"However, there were few studies with the primary aim to distinguish contraction signals and non-contraction signals using the characteristics extracted from the EHG." (research absence)*
> *"Despite the extensive use of the grout in most base plate connections, it has received limited attention [1]." (despite + neglect)*

Move 3: Occupying the Niche

This move announces what the present study does — the contribution.

Key finding: Engineering writers show the opposite preference from the humanities. Passive and third-person forms dominate: *is proposed / is presented* (6.3/10k) outweighs first-person *we present* (2.4/10k) by nearly 3:1. The distinctive engineering construction *the proposed method* (3.6/10k) frames the contribution as a thing, not an action.

Corpus examples:

> *"This paper presents a framework and an associated decision support tool specifically focused on waste heat recovery." (*third-person, system announcement*)*
> *"This paper presents a novel real-time interpolator which is applicable to not only NURBS but also general parametric curves." (*novelty claim*)*
> *"In this paper, a highly parallelizable Newton-type method is proposed." (*passive *proposal)*

Usage	Count	Per 10k	Function
in this paper / in this study / in this work	156	7.4	Locating the contribution
is proposed / is presented / is developed	133	6.3	Passive proposal announcement
the aim/purpose/objective/goal of	78	3.7	Formal aim statement
the proposed [method/model/approach]	75	3.6	Framing the contribution
we [present/propose/develo p/investigate]	50	2.4	First-person announcement
this paper/study + [presents/proposes...]	45	2.1	Third-person announcement
a novel / a new	43	2.1	Novelty claim
the main contribution	5	0.2	Explicit contribution statement
Total Move 3	585	27.9	

The CARS Ratio: Engineering vs. Humanities

The engineering Introduction is dominated by Move 2 to a far greater degree than in the humanities. The ratio is approximately **1 : 9 : 4** (Territory : Niche : Occupying). Engineering writers spend very little space claiming that their topic is important. Instead, they invest heavily in the problem: what is wrong, what is lacking, what is needed, and why current approaches fall short.

Alternative Niche Strategies

The standard CARS model treats the niche as a gap in the literature. In engineering, the niche takes several distinct forms.

Key finding: Practical need is the most common strategy (17.5/10k). Extension/improvement is second (13.5/10k). Standard research absence is the *least* common approach in engineering

(3.4/10k). The humanities writer says: "Nobody has examined X." The engineering writer says: "Existing approaches to X suffer from Y. We propose Z, which overcomes this limitation."

Niche strategy	Total per 10k	Character
Practical Need	17.5/10k	"Something better is needed"
Extension/Improvement	13.5/10k	"We make it better"
Technical Limitation	11.3/10k	"Current methods fail"
Research Absence	3.4/10k	"Nobody has studied this"

STRATEGY A: TECHNICAL LIMITATION (11.3/10K)

The writer identifies specific deficiencies in existing methods, models, or systems: *complex*, *cannot*, *suffer from*, *drawback*, *inaccurate*, *at the expense of*.

Corpus examples:

> *"However, existing algorithms suffer from the necessity of optimisation over spaces of dynamic processes: each cost function evaluation requires numerical simulation of a system model." (suffer from* + specific deficiency)
>
> *"This technique of accommodating the external disturbance in the LQ problem by minimizing its effect has a drawback: it does not guarantee the minimization of the net costs associated with control-efforts and state-deviations."* (drawback + what it cannot guarantee)
>
> *"Due to its complex structure, the cutting temperature characteristics of CFRP cannot be determined using existing protocols for metal cutting." (cannot* + specific inability)
>
> *"For micro-/nano manipulation, a conventional major challenge is the trade-off between high rigidity, large magnification, high-precision tracking, and high-accuracy positioning."* (trade-off between competing requirements)
>
> *"A drawback of this approach is that consensus itself converges asymptotically."* (concise technical limitation)

Writer's note: Technical limitation is the most characteristic engineering niche strategy. It names the problem concretely: not "more research is needed" but "this method is computationally expensive" or "this approach suffers from poor accuracy under condition X." The specificity of the limitation directly motivates the specificity of the proposed solution.

STRATEGY B: PRACTICAL NEED (17.5/10K)

The writer justifies the study through a real-world or engineering need: *to improve, is essential, need for.*

Corpus examples:

> *"In order to reduce the design costs and to maximize the acoustic performance of aerospace structures, a robust and mature prediction of interior noise levels is required." (in order to + is required)*
> *"Accurate understanding and modelling of this phenomenon is crucial for a reliable assessment of the effects of thermal loads on concrete structures." (is crucial for)*
> *"The increasing demand for structures made of CFRP in aerospace industry is enhancing the development of more eco-efficient manufacturing." (increasing demand for)*
> *"Measuring and characterising surface topography is critical for the product inspection and quality control." (is critical for)*

Writer's note: Practical need works differently from research absence. Rather than saying "nobody has studied X," it says "the engineering challenge of X requires a better solution." The key construction is the purpose infinitive: *to improve, to reduce, to achieve.*

STRATEGY C: RESEARCH ABSENCE (3.4/10K)

The writer notes that something has not been studied: *has not been investigated, few studies, overlooked.* This is the least common engineering strategy.

Corpus examples:

> *"Despite the extensive use of the grout in most base plate connections, it has received limited attention [1]." (despite* + limited attention: combines practical importance with neglect)
> *"Surprisingly, only a few studies have been conducted in this area for both subsonic and supersonic speeds." (few studies* - with the evaluative adverb *surprisingly)*

"Although currently the use of SFRC has become popular in practice, the structural behaviour of fastening systems in SFRC has not been fully investigated." (although + has not been investigated)
"So far only limited research has been done on the calibration of finite element models for half-joint applications." (limited research)
"The drill-exit temperature characteristics in the drilling of CFRP and the associated effects on drilling qualities are yet to be revealed." (yet to be - incompleteness)

Writer's note: Research absence is the standard gap strategy in the humanities and social sciences, but, as we have noted, it is the least common in engineering. When used, it is typically combined with a practical motivation: not just "nobody has studied X" but "despite its importance in Y, X has received limited attention."

STRATEGY D: EXTENSION/IMPROVEMENT (13.5/10K)

The writer positions the study as extending or improving existing work: *improve, extend, contributes to.*

Corpus examples:

"In the present work this approach is extended to the case of semi-active suspensions." (extended to - applying existing method to new domain)
"This kind of high fidelity characterisation of turbines can greatly improve the accuracy of prediction of the rotor terminal speed." (improve the accuracy - performance improvement)
"This work extends the current blockage problem to strongly sheared current flow, which is a more realistic representation of a current flow in some regions in the open sea." (extends to - generalising to more realistic conditions)
"In this paper, we build on the results in Papachristodoulou and Peet (2006), where the use of SOS programs and computationally efficient methods for stability analysis of a class of linear parabolic PDEs were reported." (build on - explicitly citing prior work being extended)

Writer's note: Extension is the second most common niche strategy in engineering. The contribution is not that something is unstudied, but that existing work can be taken further: generalised to new conditions, applied to a different system, improved in accuracy or efficiency.

Combining Strategies

In practice, many engineering Introductions combine two or more niche strategies. Common combinations:

Technical limitation + Practical need: *"Existing methods are computationally expensive* [limitation], *making them impractical for real-time applications* [need]. *This paper proposes a computationally efficient alternative."*

Research absence + Practical need: *"Despite the extensive use of X in industry* [need], *limited research has examined Y* [absence]."

Technical limitation + Extension: *"Current approaches suffer from poor accuracy under Z conditions* [limitation]. *We extend the method of* [Author] *by incorporating W* [extension]."

Implications for Writing

Move 2 Is the Engine

In engineering writing, the niche move is not merely a hinge between territory and contribution, it is the largest single rhetorical component by a wide margin (64.1/10k). If you skip or abbreviate Move 2, your Introduction will be lacking in justification. The extended niche discussion reflects the need to position your method or model against the specific limitations of existing approaches.

The "However" Pivot

The word *however* (13.8/10k) is the primary structural pivot in the engineering Introduction, appearing roughly once every 700 words. It marks the turn from what exists to what is lacking. If your Introduction does not contain a clear *however* moment, it probably lacks the turn that motivates your study.

Passive Over Personal

Engineering Introductions favour passive and third-person constructions for announcing the study. *A method is proposed* and *This paper presents* are far more common than *We propose*. This is a disciplinary convention, not a universal rule, but the corpus norm is clear.

Cite Limitations, Not Just Achievements

The most effective engineering Introductions review prior work not just to show what has been done, but to identify what each cited work *cannot* do. This creates a natural transition to the niche: "[Author] *proposed X, but this approach is limited to…* [Author2] *extended this to* Y, *but at the expense of… In this paper, we address both limitations by…*"

Unit 4: Presenting Mathematical and Theoretical Development

Engineering Corpus — Full corpus (163 articles, ~610,000 words). Modelling/Theory/Simulation content appears in 49.7% of articles.

Overview

This unit addresses the rhetorical work that engineering writers do when they develop a mathematical model, derive governing equations, state assumptions, define variables, and justify modelling choices. This appears in nearly half the corpus: in sections with headings like *System Model*, *Mathematical Formulation*, *Governing Equations*, *Theoretical Background*, or *Finite Element Model.*

The unit describes six characteristic moves. Unlike the Introduction, where moves follow a predictable sequence, the moves in a modelling section are typically interwoven: an assumption is stated, an equation is introduced, its variables are defined, a derivation step follows, a choice is justified, and the cycle repeats. The sequence is dictated by the logic of the mathematical development, not by a fixed template.

Before You Write: Conceptualising Your Theoretical Section

Step 1: State Your Purpose

Write one sentence: What is the mathematical or theoretical development in this section trying to achieve?

Step 2: List Your Assumptions

Write down every simplifying assumption you make. For each, ask: *Why is this assumption reasonable?* and *What would change if I relaxed it?*

Step 3: Map Your Derivation

Write the logical sequence of your development: What comes first?

What depends on what? Each step should lead naturally to the next.

Step 4: Identify Your Key Equations

Which equations are the main results? These are the equations that need the most careful introduction and the fullest variable definitions.

Move 1: Stating Assumptions (6.0/10k)

Engineering models always involve simplifications. This move makes those simplifications explicit.

Usage	Count	Per 10k	Function
assumption(s)	184	3.0	Naming assumptions explicitly
assuming (that)	48	0.8	Participial assumption
was/is assumed to be	38	0.6	Passive assumption
is considered (to be/as)	33	0.5	Treating as
it is assumed (that)	32	0.5	Impersonal assumption
is neglected / is ignored / is negligible	23	0.4	Dismissing a factor
for simplicity	5	0.1	Justifying simplification
Total Move 1	365	6.0	

Key finding: Assumption-stating is a pervasive move. The word *assumption* alone appears 184 times. The dismissal construction *is neglected / is negligible* (0.4/10k) serves a specific function: it justifies omitting a term or factor from the model.

Corpus examples:

> *"It is assumed in the LQ problem that all of the inputs given to the system are control inputs, which means that each of the input can be*

manipulated by the controller." (it is assumed + consequence)
"The temperature of the instrument is assumed to be 20 °C, with a standard uncertainty of 0.5 °C." (is assumed to be + specific value)
"For simplicity, due to symmetry the design of the structure is reduced to that of one planar truss." (for simplicity + what is simplified)
"In many studies [4-6], the effect of grout is neglected by not including it in the test and/or by placing the base plate in direct contact with the concrete pedestal." (is neglected + how)
"The cost of fabrication for the passive and the adaptive structure is assumed to be identical because both structures require fabrication of an equal number of joints and elements." (is assumed to be + because justification)

Writer's note: The most effective assumption statements combine three elements: (1) the assumption itself, (2) the reason for it, and (3) the consequence. The corpus shows that the best writers justify their assumptions rather than merely listing them.

Move 2: Introducing and Presenting Equations (5.4/10k)

This move introduces a mathematical expression and connects it to the argument.

Key finding: The dominant construction is *leads to* (2.1/10k), which reflects the derivational logic of engineering writing: one step *leads to* the next. The construction *yields* (0.8/10k) serves a similar function. The more static constructions: *is given by*, *is defined as* are used for presenting established relationships rather than deriving new ones.

Corpus examples:

"Substituting the kinematic and potential energies into the Lagrange equation yields..." (yields - result of derivation)
"Solubility can be defined as S = msolute/msolution, where msolute is the mass of the solute." (is defined as + where clause)

"This force can be expressed as a function of the slot's position and width, in order to provide quantitative estimates for the system." (can be expressed as + purpose)

Usage	Count	Per 10k	Function
leads to	129	2.1	Result of derivation step
is calculated / is computed	53	0.9	Computational definition
yields	48	0.8	Mathematical consequence
can be obtained	26	0.4	Derivable result
is defined as	24	0.4	Formal definition
is given by	12	0.2	Standard equation introduction
is determined by/from	12	0.2	Determination
can be expressed as	4	0.1	Alternative formulation
Total Move 2	330	5.4	

Writer's note: Choose your verb to match the rhetorical function. Use *is defined as* when establishing a term for the first time. Use *is given by* for a known relationship. Use *yields* or *leads to* when the equation follows from a preceding derivation step.

Move 3: Defining Variables and Terms (8.4/10k)

After an equation, engineering writers must define every variable.

Key finding: The single most frequent construction is *respectively* (6.4/10k). This word appears once every 156 words across the entire corpus, mapping a series of terms to a series of definitions in parallel. The *where X is the…* construction is the standard post-equation variable definition.

Corpus examples:

> *"LCOE for the plant using SC as a power block is 0.0947 $/KWh which is lower than the GC and OC by 31.82% and 48.8%, respectively."* (*respectively* - mapping values to items)
> *"Where ρc is the carrier phase density, Ucross is the gas speed averaged across the tube volume; D is the horizontal pipe diameter and μc is the viscosity of the gaseous phase."* (*where* - multiple definitions in sequence)
> *"This FDS conjunction processing and manoeuvre rate is denoted by the red circle superimposed on the JSpOC red trend line shown in Fig. 49."* (*denoted by* - symbol naming)

Writer's note: The *where* clause is the standard construction. It typically follows the equation immediately, listing each variable in turn. Using *respectively* incorrectly where the mapping is ambiguous is a common error. Ensure the two lists are in the same order and of the same length. Also make sure you are not using *respectively* when there is a single list, not a parallel one.

Move 4: Signposting Derivations (27.1/10k)

This move guides the reader through the steps of a mathematical derivation.

Usage	Count	Per 10k	Function
hence / thus / therefore	1,138	18.7	Logical consequence
by using / by applying	151	2.5	Method of derivation
simplifying	129	2.1	Algebraic reduction
applying	111	1.8	Applying a principle
differentiating	43	0.7	Calculus operation
combining	36	0.6	Merging equations
substituting	20	0.3	Replacing a variable
integrating	18	0.3	Calculus operation
Total Move 4	~1,653	27.1	

Key finding: Derivation signposting is dominated by *hence*, *thus*, and *therefore* (combined 18.7/10k, roughly once every 54 words). The gerund constructions, *substituting*, *combining*, *applying*, name the specific mathematical operation.

Corpus examples:

> *"Substituting the kinematic and potential energies into the Lagrange equation yields..." (substituting + yields* - operation and result)
> *"Since Newton's model does not hold, it follows that our oscillator is no longer isochronous." (it follows that* - formal logical consequence)
> *"Optimization has been achieved by using a multi-variable steady state optimizer to find the optimum operating conditions." (by using* - method)

Writer's note: Every derivation step needs a signpost. The pattern is: **[operation]** + **[object]** + **[result verb]** + **[equation]**. For example: *"Substituting Eq. (3) into Eq. (5) yields Eq. (6)."* The gerund form is preferred over the finite verb by a wide margin.

Move 5: Justifying Modelling Choices (66.8/10k)

This move explains *why* a particular approach, assumption, method, or parameter was chosen. It is by far the densest move in the modelling section.

Usage	Count	Per 10k	Function
based on / according to	1,041	17.1	Grounding in authority
due to / owing to	923	15.2	Causal explanation
because / since	877	14.4	Direct reason
to ensure / to guarantee / to achieve	523	8.6	Purpose
in order to / so as to	393	6.5	Purpose (formal)
is adopted / is employed / is chosen	312	5.1	Passive method selection
Total Move 5	~4,069	66.8	

Key finding: Justification is the single most important rhetorical function in the engineering modelling section. At 66.8/10k, it occurs roughly once every 15 words. The three dominant constructions are *based on* (17.1/10k), *due to* (15.2/10k), and *because/since* (14.4/10k).

Corpus examples:

> *"A time-domain optimal control approach is adopted in which ride and handling are modelled as exogenous disturbances acting on the vehicle."* (*is adopted* + description)
> *"In this approach, a flux-based Annual Collision Probability (ACP) approach is employed to estimate collision likelihood."* (*is employed* + purpose)
> *"In order to generate the flow solutions for the different cases, two different computational domains are defined at each axial displacement location."* (*in order to* + action)

Writer's note: Engineering readers expect every choice to be justified. If you select a particular mesh density, turbulence model, boundary condition, or solution algorithm, the reader will ask *why*. The justification need not be lengthy — a subordinate clause with *because*, *due to*, or *in order to* is often sufficient — but it must be present.

Move 6: Describing the Model or System (6.7/10k)

This move presents the physical or computational system being modelled.

Corpus examples:

> *"The Efficient Cruise Control (ECC) design consists of two controllers as shown in Fig. 2."* (*consists of* - structural composition)
> *"The preceding vehicle's initial speed remains the same and the initial inter-vehicular distance is governed by (3)."* (*is governed by* - referencing an equation)
> *"One can introduce averaged boundary conditions and perform steady-state simulations, which is even 7 times less time-consuming."* (boundary conditions — constraints)

Usage	Count	Per 10k	Function
consists of / comprises / is composed of	148	2.4	Structural description
the model / the system [+ verb]	105	1.7	Subject-led description
boundary condition(s)	84	1.4	Specifying constraints
initial condition(s)	42	0.7	Specifying starting state
is modelled as / is treated as	11	0.2	Classification
is governed by	10	0.2	Naming governing physics
governing equation(s)	9	0.1	Naming the equations
Total Move 6	409	6.7	

Writer's note: Begin with the big picture: what the system is and what it consists of, then zoom in to governing equations, boundary conditions, and parameter values. The construction *is governed by* is particularly useful for connecting the physical description to the mathematical formulation.

The Move Cycle

Unlike the Introduction, the modelling section cycles through its moves repeatedly. A typical passage:

Move 6 (Describe): *"The system consists of a cantilever beam with a piezoelectric actuator bonded to the upper surface."*

Move 1 (Assume): *"The beam is assumed to be an Euler-Bernoulli beam, and shear deformation is neglected."*

Move 2 (Introduce): *"The equation of motion is given by..."*

Move 3 (Define): *"...where E is Young's modulus, I is the second moment* *of area, and ρ is the density."*

Move 5 (Justify): *"The Euler-Bernoulli assumption is valid because the beam's length-to-thickness ratio exceeds 20."*

Move 4 (Derive): *"Substituting the boundary conditions into Eq. (3) yields..."*

This cycle — describe, assume, formulate, define, justify, derive — is the backbone of the engineering modelling section. It may repeat several times as the model is built up layer by layer.

Implications for Writing

Justify Everything

Justification language (66.8/10k) is far denser than any other move. Every assumption, every parameter choice, every method selection should be accompanied by a reason.

Define Every Variable

After every equation, define every new symbol. The *where* clause is the standard construction. Do not assume the reader will recognise your notation.

Signpost Every Step

In a derivation, the reader should always know what operation you are performing and why. The gerund form (*substituting*, *combining*, *simplifying*) is the standard way to name the operation.

State Assumptions Explicitly

Do not bury assumptions in your derivation. State them clearly, using *it is assumed that*, *for simplicity*, or *is assumed to be*. Where possible, justify the assumption and state its consequence.

Unit 5: Describing Experimental and Computational Methods

Engineering Corpus — Full corpus (163 articles, ~610,000 words)

Overview

The Methods section describes what was done and why: the physical experiments conducted, the computational simulations run, or the procedures followed. It is the most procedural section of the paper, dominated by the past tense and the passive voice.

Engineering methods sections are more diverse than their humanities equivalents. They may describe a physical experimental setup, a computational simulation environment, or both (in Path A papers). They often appear under technical headings such as *Experimental Setup*, *Computational Domain and Boundary Conditions*, *Test Procedures* rather than as a generic "Methods" section.

This unit covers five moves. Move 1 (equipment and materials) and Move 3 (procedural description) are shared with experimental sciences generally. Move 2 (parameter specification) and Move 5 (computational setup) are distinctive to engineering. In addition, there is Move 4 (justifying choices), which is found more generally.

Before You Write: Conceptualising Your Methods Section

Step 1: Name Your Approach

Write one sentence: What kind of experimental or computational work did you do?

Step 2: Describe Your Setup

What equipment, software, or facilities did you use? What materials or specimens were tested?

Step 3: List Your Parameters

Write down every parameter that matters. For each, note the value you used and *why*.

Step 4: Map Your Procedure

Write the steps in chronological order. For each, ask: *Could another researcher replicate this?*

Step 5: Check for Justification

For each major choice respond to why this software, why this mesh density, why this test configuration.

Move 1: Specifying Equipment, Software, and Materials (49.1/10k)

This move names the tools, instruments, software, and materials used.

Usage	Count	Per 10k	Function
specimen(s) / sample(s) / material(s)	1,731	28.4	Naming what was studied
was/were used	308	5.1	General tool specification
was/were carried out / performed / conducted	236	3.9	Announcing the procedure
FEM / FEA / finite element	235	3.9	Naming computational method
was/were measured / recorded / monitored	164	2.7	Measurement specification
CFD	131	2.2	Naming simulation type
Software names (MATLAB, ANSYS, ABAQUS, etc.)	129	2.1	Naming software tools
Total Move 1	~2,991	49.1	

Key finding: Material and specimen description dominates at 28.4/10k. The passive procedural announcement (*was carried out, was performed* at 3.9/10k) is the standard way to open a methods description. Named software tools appear at 2.1/10k; the most frequent are MATLAB, ANSYS, ABAQUS, FLUENT, COMSOL, and LabVIEW.

Corpus examples:

> *"Finite element plain strain analysis was carried out using GEO5 FEM software, Fine Ltd (2018)." (was carried out using* - procedure + tool)
> *"The joint overlap elongation was measured using a laser extensometer to exclude the compliance of the machine." (was measured using -* instrument + purpose)
> *"The waste heat survey was conducted using two Temperature Data Loggers, inlet flow rate was measured using standard flow metre, and the outlet flow rate was calculated theoretically."* (multiple instruments in sequence)
> *"The controller is implemented in MATLAB and a Tesla S vehicle model from IPG CarMaker is used to simulate an EV."* (software + simulation environment)

Writer's note: The key construction is **[procedure] + was carried out/performed + using + [tool/software]**. When naming software, include the version number and the developer for replicability.

MOVE 2: SPECIFYING PARAMETERS AND CONDITIONS (23.0/10K)

This move states the values, settings, and conditions under which the work was done.

Key finding: The word *parameter(s)* alone appears 910 times (14.9/10k). The convergence study — verifying that results do not change with finer discretisation — is a distinctive engineering requirement (2.2/10k).

Usage	Count	Per 10k	Function
parameter(s)	910	14.9	Naming parameters generically
was/were selected / chosen	160	2.6	Announcing parameter selection
convergence	136	2.2	Mesh/solution convergence
mesh / grid / element size	73	1.2	Spatial discretisation
ranged from / ranging from	57	0.9	Specifying ranges
time step / timestep	41	0.7	Temporal discretisation
was/were set to / fixed at	24	0.4	Specifying exact values
Total Move 2	~1,401	23.0	

Corpus examples:

"The explicit Euler method is used for the C/GMRES method, and the number of iterations in GMRES is set to 7." (is set to - exact parameter value)
"Three PPCAs (P-841.30, from PI Ceramic GmbH) with a closed-loop stroke of 45 μm were selected to drive the platform." (were selected - manufacturer + specification)
"Maximum vertical displacements of the foundation plate ranged from 3.7 mm to 4.1 mm upwards." (ranged from - range specification)
"The mesh size and the distribution of points were decided based on previous experience with cavity flow simulation." (mesh specification + justification)

Writer's note: Be precise. Do not write *"a fine mesh was used"* rather write *"a mesh of 50,000 hexahedral elements with a maximum edge length of 0.5 mm was used."* Every parameter should have a value, a unit, and a justification.

Move 3: Describing Procedures (93.1/10k)

This move narrates what was done, step by step.

Usage	Count	Per 10k	Function
Past passive (was/were + past participle)	3,593	59.0	Procedural backbone
first / firstly / initially	806	13.2	Opening step
then / subsequently / afterwards	611	10.0	Following step
followed by / prior to / before / after	321	5.3	Relative sequencing
finally / lastly	196	3.2	Closing step
next / the next step	145	2.4	Step transition
Total	~5,672	93.1	

Key finding: The past passive at 59.0/10k is the grammatical backbone, occurring roughly once every 17 words. This is denser than the humanities figure (42.5/10k). Temporal sequencing markers (34.1/10k) create the chronological structure.

Corpus examples:

> *"The waste heat survey was conducted using two Temperature Data Loggers, inlet flow rate was measured using standard flow metre, and the outlet flow rate was calculated theoretically."* (three-step passive sequence)
> *"Firstly, the experiment on the first DLM was conducted along a single axis." (firstly* - opening step)
> *"The simulation was performed on a desktop computer with dual 2.2 GHz Intel Xeon E5-2650 V4 processors with 24 cores in total." (was performed on* - computational environment)

Writer's note: Each step is a completed action. The passive voice keeps the focus on the procedure rather than the researcher. The temporal markers (*first, then, finally*) provide the chronological scaffolding.

Move 4: Justifying Methodological Choices (16.2/10k)

Usage	Count	Per 10k	Function
in order to / so as to	393	6.5	Purpose
to ensure / to avoid / to prevent / to minimise	327	5.4	Safeguarding purpose
in accordance with / according to	231	3.8	Citing standards or authorities
was validated / was verified / was calibrated	34	0.6	Verification statement
Total Move 4	~988	16.2	

This move explains why specific equipment, parameters, or procedures were selected.

Key finding: The most characteristic engineering construction is the appeal to standards: *in accordance with / according to* (3.8/10k). Engineering methods are frequently anchored to named standards (ISO, BS, ASTM, Eurocode, ASME).

Corpus examples:

> *"SB: Standard bonds prepared according to the FM® 94 specification with temperature rate of 2 °C/min." (according to* - citing a specific standard)
> *"No advanced configurations have been proposed to the schemes to avoid any additional cost by adding more components." (to avoid* - safeguarding purpose)
> *"The use of a quasi-static model of the load forces is necessary both for effective control and to ensure good performance of the estimator." (to ensure* - purpose justification)

Writer's note: The construction **according to [standard/specification]** is the most distinctively engineering form of methodological justification. Where the humanities writer cites *following [Author] (year)*, the engineering writer cites *in accordance with ISO 4287*.

Move 5: Specifying Computational Setup (48.3/10k)

Usage	Count	Per 10k	Function
simulation(s)	800	13.1	Naming computational work
node(s) / element(s)	797	13.1	Mesh components
solver / algorithm	580	9.5	Solution method
mesh / grid	277	4.5	Spatial discretisation
discreti(s/z)ation	188	3.1	Discretisation method
iteration(s) / iterative	149	2.4	Solution process
turbulence model (RANS, LES, DES, SST)	91	1.5	Physics model selection
was/were implemented / coded	60	1.0	Code development
Total Move 5	~2,942	48.3	

This move is unique to engineering. It describes the computational environment: the solver, the mesh, the discretisation scheme, the convergence criteria, and the hardware.

Key finding: Computational setup language (48.3/10k) is nearly as dense as equipment specification (49.1/10k), reflecting the equal importance of simulation and experiment in engineering. The solver/algorithm vocabulary (9.5/10k) is particularly rich.

Corpus examples:

> *"The proposed method was implemented using the open-source toolkit ParNMPC, which can automatically generate parallel C/C++ code with OpenMP." (was implemented using* - code and toolkit)
> *"The algorithm described in the methodology section was implemented in MATLAB on a desktop PC with 2 Intel Xeon CPU E5-2687W @ 3.10 GHz processors."* (implementation + hardware)
> *"The FHOCP is first discretized using an explicit integration scheme performed in the reverse-time direction, enabling the problem to be split into linearly coupled subproblems." (discretised using* - method + consequence)

Writer's note: Specify four things: (1) the **software or code**, (2) the **discretisation** (mesh type, element count, time step), (3) the **solver settings** (convergence criteria, iteration limits), and (4) the **hardware** (processor, cores, run time).

Methods vs. Modelling: When to Use Which Unit

If your section...	Use...
Derives equations, states assumptions, defines variables	Unit 4 (Mathematical Development)
Describes physical experiments, test procedures	Unit 5, Moves 1–4
Describes computational simulations, mesh, solver	Unit 5, Move 5 (+ Unit 4 for the underlying model)
Does both (Path A papers)	Unit 4 for theory, Unit 5 for experiment and computation

The Methods Section Template

Equipment and materials (Move 1, ~49.1/10k): Name the instruments, software, materials, and specimens.

Parameters and conditions (Move 2, ~23.0/10k): State every numerical parameter, its value, and its unit.

Procedures (Move 3, ~93.1/10k): Narrate the steps in past passive, with temporal sequencing.

Justification (Move 4, ~16.2/10k): Thread reasons throughout — *in order to, to ensure, according to [standard].*

Computational setup (Move 5, ~48.3/10k): Specify solver, mesh, discretisation, convergence, and hardware.

Implications for Writing

The Passive Voice Is Standard

The past passive (59.0/10k) is not a stylistic preference; it is the grammatical norm. Use it fairly consistently for procedural description.

Cite Standards

Where the humanities writer cites scholarly precedent, the engineering writer cites standards (ISO, ASTM, BS, Eurocode) at 3.8/10k. If a standard exists for your procedure, cite it.

Be Numerically Precise

Every parameter should have a value and a unit. Every mesh should have an element count and a convergence study. Vagueness undermines replicability.

Describe Both Physical and Computational

In Path A papers, you need both a modelling section and an experimental section. In Path B papers, your "methods" section *is* the computational setup. Match the level of detail to the type of work.

Unit 6: Reporting and Interpreting Results

Engineering Corpus — Full corpus (163 articles, ~610,000 words)

Overview

The Results section presents what was found. In engineering, results are overwhelmingly communicated through figures, tables, and equations; the surrounding text directs the reader's attention to key features, reports numerical outcomes, and provides brief commentary.

Unlike the humanities, where results sections are dense with p-values and test statistics, engineering results sections are dense with **directional change language** (*increase, decrease* at 54.4/10k), **extrema** (*maximum, minimum, peak* at 24.2/10k), and **comparisons** (*higher than, compared to* at 15.5/10k combined). The Results section in engineering is less about statistical significance and more about performance, behaviour, and agreement with predictions.

Before You Write: Conceptualising Your Results Section

Step 1: Identify Your Main Findings

List your findings, starting with the most important. For each, write one sentence in plain language.

Step 2: Decide Your Organising Principle

By test case or configuration; from validation to new results; by parameter; or from general to specific.

Step 3: Plan Your Data Displays

For each finding: figure, table, or text? In engineering, figure references (60.0/10k in body sections) far exceed table references (10.3/10k).

Step 4: Decide What to Comment On

Brief commentary (6.6/10k) is acceptable in Results, particularly *good agreement* judgments, but save extended interpretation for the Discussion.

Move 1: Directing Attention to Data Displays (10.6/10k)

Engineering results are communicated primarily through figures and tables. This move directs the reader's attention.

Usage	Count	Per 10k	Function
Fig./Figure N shows/presents/illustrates	350	5.7	Figure as grammatical subject
is/are shown/presented/plotted in Fig./Table	104	1.7	Finding as subject, display as location
Table N shows/presents/lists	85	1.4	Table as grammatical subject
it can be seen/observed from	56	0.9	Impersonal observation
as shown/seen/presented in Fig./Table	28	0.5	Parenthetical back-reference
see Fig./Table	25	0.4	Directive reference
Total Move 1	648	10.6	

Key finding: The figure-as-subject construction (*Fig. N shows…* at 5.7/10k) is the dominant way to open a results paragraph. The impersonal *it can be seen* (0.9/10k) is a distinctively engineering construction that removes all human agency from the observation.

Corpus examples:

> *"Fig. 9 shows the SPL evaluated at 'Head Cavities' for some section and for all the 5 investigated configurations." (Fig. N shows* - figure as subject)
> *"Table 3 shows the electric vehicle's energy consumption with and without the proposed ECC under these initial conditions." (Table N shows* - table as subject)
> *"Table 8 shows that the tracking accuracy on a 50 μm-diameter circular trajectory was less than 5 nm and the homing accuracy was less than 1 nm." (Table N shows that* - introducing a finding)
> *"From these figures it can be seen that the ensemble of ascending nodes occupied by the inclined debris fragments is centred at 0° spanning ±60°." (it can be seen* - impersonal observation)

Writer's note: Choose your construction to control emphasis. *Fig. 5 shows X* foregrounds the figure. *X is shown in Fig. 5* foregrounds the finding. *It can be seen from Fig. 5 that X* creates an observation frame. Vary them to avoid monotony.

Move 2: Reporting Findings (134.3/10k)

The dominant move: presenting specific outcomes with numerical evidence.

Key finding: Where the humanities rely on p-values (56.9/10k), engineering relies on **directional change** (*increase/decrease* at 54.4/10k) and **extrema** (*maximum/minimum/peak* at 24.2/10k). Engineering Results describe how things change, how big they get, and how they compare.

Corpus examples:

> *"LCOE for the plant using SC as a power block is 0.0947 $/KWh which is lower than the GC and OC by 31.82% and 48.8%, respectively." (comparative + respectively)*
> *"Results indicate that a collision is likely to occur every 4 years for one satellite out of the entire GEO active satellite population against a 1 cm RSO catalogue." (results indicate* - quantified finding)

"The proposed ECC reduces the electric vehicle's energy consumption by approximately 23.56%." (directional change + approximate quantification)

Usage	Count	Per 10k	Function
increase/decrease/rise/decline/drop/reduce	3,312	54.4	Directional change
maximum/minimum/peak/optimal	1,475	24.2	Extrema and optima
shows/showed that / indicates that	783	12.9	Reporting verb + that-clause
significant(ly)	677	11.1	Marking significance
higher/lower/greater/less than	548	9.0	Comparative reporting
compared to/with	395	6.5	Explicit comparison
the effect/influence/impact of	391	6.4	Effect-framing
approximately / roughly / about [N]	284	4.7	Approximate quantification
results show/indicate/suggest	229	3.8	Results as subject
Total Move 2	~8,180	134.3	

Writer's note: The characteristic pattern is: **[what changed] + [direction] + [by how much] + [compared to what]**. For example: *"The heat transfer coefficient increased by 15% compared to the baseline configuration."* Always include direction, magnitude, unit, and comparison basis.

Move 3: Commenting on Findings (6.6/10k)

Brief interpretive commentary, anchored to data.

Key finding: The distinctively engineering construction is **good agreement** (1.0/10k) — used to judge the match between model predictions and experimental data. It is a validation marker embedded within the Results section.

Usage	Count	Per 10k	Function
this suggests/indicates/implies /confirms	107	1.8	Brief interpretation
consistent with / in line with / in agreement with	88	1.4	Anchoring to expectations
this is due to / because / attributed to	71	1.2	Brief causal explanation
good/reasonable/excellent agreement	63	1.0	Validation commentary
as expected / not surprisingly / notably	53	0.9	Evaluative comment
may be / might be / could be [explanation]	21	0.3	Hedged explanation
Total Move 3	403	6.6	

Corpus examples:

> *"For the current simulation the measured and calculated turbulence intensity show a good agreement at the various mass loading." (good agreement* - validation)
> *"This is because the proposed ECC is only active within a certain distance, in this work 500 m, of the upcoming traffic signal." (this is because* - brief causal explanation)
> *"As expected, flow reduction is observed when the incident sheared flow encounters the porous tower." (as expected* - confirming predictions)
> *"This is because at very low expansion ratios the turbine is not producing significant amounts of work at any non-dimensional speed line." (this is because* - explaining a mechanism)

Writer's note: Keep Results commentary brief. *This is because* and *this is due to* offer quick causal explanation. *Good agreement* signals validation. Save extended interpretation for the Discussion (Unit 8).

Move 4: Transitioning Between Findings (62.2/10k)

Usage	Count	Per 10k	Function
first/second/third/finally	1,455	23.9	Sequential ordering
however / on the other hand / in contrast	1,139	18.7	Contrastive transition
moreover/furthermore/additionally/in addition	642	10.5	Additive connection
with respect to / in terms of / regarding	468	7.7	Topic-shift markers
similarly / likewise	81	1.3	Parallel transition
Total Move 4	~3,785	62.2	

Key finding: Engineering uses *however* / contrastive markers (18.7/10k) far more heavily than the humanities, reflecting the comparative nature of engineering results. The construction *in terms of* (7.7/10k) shifts the dimension of comparison: *"In terms of accuracy, Method A is superior; in terms of computational cost, Method B is preferred."*

Corpus examples:

> *"In terms of ride comfort, the rms value of the sum of the body accelerations was increased by approximately 5% compared to the proposed observer." (in terms of* - dimension of comparison)
> *"In this case it has been considered only the Head Cavity of the section III because it appears to be the most critical in terms of internal SPL." (in terms of* - narrowing focus)

The Results Cycle

For each major finding, the moves cycle:

Direct (Move 1): *"Fig. 5 shows the variation of displacement with load for all three configurations."*

Report (Move 2): *"The maximum displacement for Configuration A was 12.3 mm, compared to 8.7 mm for Configuration B, a reduction of 29%."*

Comment (Move 3): *"This reduction is consistent with the increased stiffness predicted by the analytical model."*

Transition (Move 4): *"In terms of energy absorption, however, Configuration A outperformed Configuration B…"*

Then repeat for the next finding.

Results vs. Discussion

Feature	Results	Discussion
Numerical detail	Dense (134.3/10k)	Less dense
Figure/table references	High (10.6/10k)	Lower
"Good agreement"	1.0/10k (brief validation)	Extended comparison
Causal explanation	Brief (this is because 1.2/10k)	Extended (may be attributed to)
Hedging	Lower	Higher
Comparison with literature	Brief anchoring (1.4/10k)	Extended discussion

Implications for Writing

Lead with Figures

The construction *Fig. N shows…* (5.7/10k) is the standard opening for a results paragraph. Plan your figures first, then write the text around them.

Quantify Everything

Always state the direction, magnitude, unit, and comparison basis. Vague reports (*"the performance improved"*) are insufficient.

Use "Good Agreement" for Validation

The construction *good agreement* (1.0/10k) signals the match between prediction and measurement. It belongs equally in Unit 7 (Validation).

Vary Your Display References

Alternate between *Fig. N shows…* (figure as subject), *X is shown in Fig. N* (finding as subject), and *It can be seen from Fig. N that…* (impersonal). Monotonous repetition is a common weakness.

Unit 7: The Validation Move

Engineering Corpus — Full corpus (163 articles, ~610,000 words)

Overview

Validation, i.e., demonstrating that a model, simulation, or method produces results that agree with independent evidence, is one of the defining rhetorical acts of engineering writing.

Validation language appears across multiple sections: in the Introduction (promising validation), in the Results (reporting agreement), and in the Conclusion (claiming validation was achieved). It constitutes a recurring rhetorical move with four characteristic steps: announce the validation purpose, compare with reference data, judge the quality of agreement, and explain any discrepancies.

The corpus reveals a combined validation vocabulary density of approximately **47.2/10k** — making it one of the densest rhetorical functions in engineering writing.

Before You Write: Conceptualising Your Validation

Step 1: Identify Your Reference

What are you validating against? Experimental measurements, analytical solutions, numerical benchmarks, standards, or field data?

Step 2: Choose Your Metrics

How will you measure agreement? Visual comparison, percentage error, RMS error, correlation coefficient?

Step 3: Decide Your Threshold

What counts as "good agreement"? State your threshold explicitly or cite a standard that defines it.

Step 4: Anticipate Discrepancies

Where do you expect disagreement? Having explanations ready makes the discrepancy discussion more convincing.

Move 1: Announcing the Validation Purpose (13.0/10k)

Usage	Count	Per 10k	Function
validated / validation	277	4.5	Naming validation explicitly
verified / verification	210	3.4	Naming verification
calibrated / calibration	179	2.9	Naming calibration
benchmark(ing)	72	1.2	Naming benchmark comparison
to validate / to verify	52	0.9	Purpose infinitive
Total Move 1	790	13.0	

Key finding: *Validation* (4.5/10k) and *verification* (3.4/10k) are not synonyms. Validation asks "Are we solving the right equations?" (model vs. reality). Verification asks "Are we solving the equations right?" (numerical accuracy). *Calibration* (2.9/10k) involves adjusting parameters to fit data. *Benchmarking* (1.2/10k) involves comparing against a known reference solution.

Corpus examples:

> *"Model validation of finite element models for the low frequency range is mainly based on the correlation of eigenvalues and eigenvectors from experiments and simulations." (validation* - FE model against experimental data)
> *"Experimental results validate the proposed design and modelling methods." (validate* - experiment confirms model)
> *"An Adaptive Cruise Control (ACC) is used to benchmark the proposed ECC." (benchmark* - comparison with existing method)

"In order to benchmark the proposed ECC from Section 3.1, its performance is compared with that of the ACC from Section 4.1." (benchmark + compared with)

Writer's note: Use the correct term. *Validate* against physical reality. *Verify* for numerical accuracy. *Calibrate* for parameter fitting. *Benchmark* against reference solutions.

Move 2: Comparing with Reference Data (14.7/10k)

Usage	Count	Per 10k	Function
compared with / to / comparison with	565	9.3	General comparison
experimental data / results / measured values	182	3.0	Naming experimental reference
FEM / FEA / numerical results	72	1.2	Naming numerical reference
analytical solution / results / closed-form	64	1.1	Naming analytical reference
published / literature / reference data	12	0.2	Naming published reference
Total Move 2	895	14.7	

Key finding: The reference data is most often experimental (3.0/10k), followed by numerical (1.2/10k) and analytical (1.1/10k). This hierarchy reflects the engineering disciplinary value: physical measurement is the gold standard for validation.

Corpus examples:

"FE models of all six specimens were constructed and their predictions are compared to the experimentally recorded behaviour in Fig. 8." (FE predictions compared to experimental data)
"Various compliance/stiffness calculation equations for circular flexure hinges were compared with FEA results." (analytical compared with numerical)
"Average water demands for cultivated crops were calculated for the period of 2000-2012 and compared with published literature as shown

in Table 5." (calculated compared with published)
"The local behaviour of the assemblages was compared to the experimental test in order to verify the actual response of the connection was modelled accurately." (compared to + in order to verify)

Writer's note: The pattern is: **[your results] + compared with/to + [reference data] + [in Fig./Table N]**. Always specify what output variable is being compared, against which reference, and where the reader can see the comparison.

Move 3: Judging the Quality of Agreement (18.5/10k)

Usage	Count	Per 10k	Function
accurate / accuracy	656	10.8	General accuracy claim
error / discrepancy / deviation [+ value]	334	5.5	Quantified disagreement
good/reasonable/excellent/close agreement	63	1.0	Qualitative judgment
overestimates / underestimates	43	0.7	Directional disagreement
agree well / in good agreement	32	0.5	Agreement phrase
Total Move 3	~1,128	18.5	

Key finding: Agreement judgment operates at two levels. The **qualitative** level uses graded adjectives: *excellent* > *good* > *reasonable* > *fair*. The **quantitative** level uses error metrics (5.5/10k). The best validation passages combine both. The directional markers *overestimates/underestimates* (0.7/10k) are distinctively engineering — critical for design applications where conservative vs. non-conservative predictions have different safety implications.

The agreement scale:

Excellent/very good agreement: error < ~5%, nearly indistinguishable on plots

Good agreement: error < ~10%, trends well captured, minor quantitative differences

Reasonable/fair agreement: error < ~20%, trends captured, noticeable quantitative differences

Qualitative agreement: trends captured, quantitative match poor

Corpus examples:

> *"For the current simulation the measured and calculated turbulence intensity show a good agreement at the various mass loading."* (*good agreement* - qualitative)
> *"Reasonably good agreement between the two predictions suggests there is relatively small vertical interaction inside the porous tower."* (*reasonably good* - hedged judgment + interpretation)
> *"Despite the fact that the WSS values determined in CFD simulations may be overestimated by the assumption of rigid walls, the relative order of magnitude of the WSS is retained."* (*overestimated* + but acceptable)
> *"The blood flow rate is always overestimated due to the way of calculations in clinical examinations."* (*overestimated* - directional + causal)

Writer's note: Do not stop at *"good agreement was observed."* Support the qualitative judgment with numbers: *"Good agreement was observed, with an RMS error of 3.2% across all test cases."*

Move 4: Explaining Discrepancies (1.0/10k)

Key finding: The dominant construction is the hedged explanation: *may be attributed to*, *may be due to*, *could be explained by*. The hedging is deliberate — the writer cannot prove why the model disagrees, only suggest possible reasons.

Corpus examples:

> *"The quieter cavity flow for the store at carriage may be caused by the shear layer that is seen not to go deep in the cavity."* (*may be caused by* - hedged mechanism)

"It may be due to limitation of the aeroelasticity to the fins and the small displacements of the fins." (may be due to - model limitation)
"The fair agreement of the comparison measurements with the CMM may be caused by limitations in the typical CMM accuracy." (fair agreement + may be caused by)

Usage	Count	Per 10k	Function
may be attributed to / due to / explained by	45	0.7	Hedged causal explanation
this discrepancy / difference / deviation	13	0.2	Naming the disagreement
is likely due to / is probably due to	4	0.1	Tentative explanation
Total Move 4	63	1.0	

Writer's note: Be specific about the source: mesh resolution, simplifying assumptions, measurement uncertainty, boundary condition approximations, material property variability. Generic explanations (*"due to modelling assumptions"*) are weak. Name the specific assumption.

The Validation Sequence

The four moves typically appear in sequence:

Announce (Move 1): *"To validate the proposed model, its predictions are compared with the experimental measurements of [Author] (year)."*

Compare (Move 2): *"Fig. 8 compares the predicted and measured load–displacement curves for all six specimens."*

Judge (Move 3): *"Good agreement is observed, with a maximum error of 7.2% in the peak load prediction."*

Explain (Move 4): *"The slight overprediction at large displacements may be attributed to the elastic–perfectly plastic material model, which neglects strain hardening."*

Types of Validation in the Corpus

Validation type	What is compared	Typical reference
Model vs. experiment	Simulation output vs. physical measurement	Your own or published experiments
Method vs. benchmark	Your code output vs. known solution	Exact solutions, reference codes
Method vs. method	Your approach vs. existing approach	Competing methods in literature
Design vs. standard	Your design vs. code requirements	ISO, Eurocode, AISC, etc.
Instrument vs. reference	Your measurement vs. reference	Calibration standards, CMMs

Implications for Writing

Validate Before You Claim

Validation language (47.2/10k combined) is one of the densest rhetorical functions in the corpus. If your paper presents a model, method, or simulation, readers will expect validation. Do it before moving to parametric studies or novel predictions.

Quantify Your Agreement

Do not stop at *"good agreement was observed."* Support the qualitative judgment with numbers: error percentages, RMS values, correlation coefficients.

Explain Discrepancies Specifically

Name the specific source: which assumption, which simplification, which measurement uncertainty. Generic explanations undermine the validation.

Use the Correct Term

Validate (against physical reality), *verify* (numerical accuracy), *calibrate* (parameter fitting), *benchmark* (against reference solutions). Using these terms interchangeably suggests unfamiliarity with the validation framework.

Unit 8: Rhetorical Moves in the Discussion

Engineering Corpus — Full corpus (163 articles, ~610,000 words)

Overview

The Discussion is where you interpret your results: what do they mean, how do they compare with others' work, and what are the practical consequences? In engineering, the Discussion is often merged with Results or distributed across sections. Regardless of placement, the rhetorical moves are consistent.

The engineering Discussion has particular characteristics. First, **hedging density is low**, e.g., modals at 22.0/10k. Second, **practical implications are more prominent** (13.2/10k). Third, **explanations are more mechanistic**: engineering writers explain through physical mechanisms, not theoretical frameworks.

Before You Write: Conceptualising Your Discussion

Step 1: Restate Findings in Conceptual Terms

Translate numerical results into conceptual claims. Results: *"Energy consumption decreased by 23.56%."* Discussion: *"The proposed method achieves substantial energy savings."*

Step 2: Connect to the Literature

Does this agree with, disagree with, or extend previous work? Which studies?

Step 3: Offer Physical Explanations

What mechanism explains this? Why does the system behave this way?

Step 4: Identify Limitations

What simplifications affect validity? What parameters were not varied?

Step 5: Articulate Practical Implications

What should engineers, designers, or practitioners do differently?

Move 1: Restating Findings (8.7/10k)

Usage	Count	Per 10k	Function
results show/indicate/suggest/demonstrate	230	3.8	Third-person restatement
it was/has been shown/demonstrated/found	164	2.7	Passive restatement
as shown/discussed/mentioned above	83	1.4	Back-reference
the proposed [method/model] + verb	53	0.9	Foregrounding contribution
Total Move 1	530	8.7	

Key finding: Engineering writers predominantly use passive or third-person constructions. The construction *the proposed [method] + verb* (0.9/10k) is distinctively engineering — keeping the method as the agent.

Corpus examples:

> *"It was found that the bond area reduction by approx. 64% will result in less than 50% failure load reduction."* (*it was found* - impersonal passive restatement)
> *"The proposed ECC reduces the electric vehicle's energy consumption by approximately 25.64% in the simulations performed."* (*the proposed + reduces* - method as agent)
> *"It has been shown that with careful technique and processing the value of parameters obtained using XCT are remarkably similar to*

those obtained using conventional optical surface texture measurement techniques." (it has been shown - passive with interpretation)

Move 2: Comparing with Literature (29.3/10k)

Usage	Count	Per 10k	Function
supports / confirms / corroborates	1,192	19.6	Showing agreement
[Ref] found/reported/showed/observed	154	2.5	Citing prior findings
similar to / comparable to	129	2.1	Drawing parallels
previous/prior/existing studies/work	110	1.8	Generic literature reference
contrary to / unlike / in contrast to	109	1.8	Noting disagreement
consistent with / in line with	88	1.4	Alignment
Total Move 2	~1,782	29.3	

Key finding: The agreement-to-disagreement ratio in engineering is approximately **12:1** (21.0/10k vs. 1.8/10k). Engineering writers overwhelmingly position findings as extending or confirming previous work. Citing by reference number rather than by author name is the convention.

Corpus examples:

> *"Similar to [1], the cost function does not aim to minimize acceleration." (similar to [Ref]* - alignment)
> *"Collectively, these new findings indicate that collision likelihood in GEO is as much as four orders of magnitude higher than previously published by other researchers."* (contrast with previous)
> *"However, NMPC, unlike linear MPC, does not have a general explicit state feedback law." (unlike* - method comparison)

MOVE 3: EXPLAINING RESULTS (6.3/10K)

Usage	Count	Per 10k	Function
suggests/indicates/implies that	255	4.2	Drawing inferences
this is because / this is due to	78	1.3	Direct causal explanation
may/might/could be due to / attributed to	37	0.6	Hedged explanation
it is possible/likely/plausible that	8	0.1	Speculative framing
Total Move 3	~385	6.3	

Key finding: Engineering explanations are quite direct and mechanistic. *This is because* (1.3/10k), a rather blunt causal statement, is far more common than *one possible explanation* (0.0/10k). Engineering writers explain through physical mechanisms, not theoretical speculation.

Corpus examples:

> *"This is because the energy from the rock bed has been significantly depleted and incoming cold air no longer can be heated again to the recorded discharge temperature." (this is because* -physical mechanism)
> *"This is because the proposed ECC is only active within a certain distance, in this work 500 m, of the upcoming traffic signal." (this is because* - design explanation)
> *"This is because at very low expansion ratios the turbine is not producing significant amounts of work at any non-dimensional speed line." (this is because* - performance)

Writer's note: The pattern is: *"[Observation]. This is because [mechanism]."* When uncertain, use *may be due to* or *can be attributed to*, but engineering writers hedge explanations less than those in other more interpretive disciplines.

Move 4: Limitations (2.9/10k)

Usage	Count	Per 10k	Function
limitation(s)	91	1.5	Naming limitations explicitly
drawback / disadvantage / shortcoming	54	0.9	Technical limitation
does not account for / does not consider	25	0.4	Scope limitation
beyond the scope	6	0.1	Scope boundary
Total Move 4	176	2.9	

Key finding: Limitation density in engineering (2.9/10k) is not very high. Engineering limitations are **technical** (drawback of the proposed method) rather than methodological (study design limitations).

Corpus examples:

> *"One of the limitations of the JSpOC model is that CA is performed using the Special Perturbations catalogue for both primary and secondary objects."* (technical limitation)
> *"A limitation of the proposed method is the calculation of remaining time after which the host vehicle can pass through the upcoming traffic signal."* (method limitation)
> *"This limitation can be mitigated by sharing the orbital ephemeris including planned manoeuvres with the JSpOC."* (limitation + mitigation)

Writer's note: Brief, specific, solution-oriented: (1) Name the limitation, (2) Explain its consequence, (3) Suggest how it could be addressed.

Move 5: Practical Implications and Recommendations (13.2/10k)

Key finding: Practical implications (13.2/10k) are quite common. The dominant construction is *enables/allows/provides* (8.0/10k). Engineering writers frame contributions as enabling practical capabilities.

Usage	Count	Per 10k	Function
enables / allows / facilitates / provides	486	8.0	Capability framing
can be used/applied/employed (for/to/in)	172	2.8	Application statement
potential (for/to/of)	96	1.6	Future potential
practical/industrial/engineering application	47	0.8	Naming the application domain
Total Move 5	801	13.2	

Corpus examples:

> *"Such an analysis can be used in a computational technology selection for optimised energy recovery and minimised financial payback of implementation." (can be used* - application)
> *"The proposed method can work as a valuable enhancement tool for machine tool calibration engineers."* (practical value)
> *"It provides a framework for predicting the dynamic response and analyzing the vibro-acoustic transfer paths of complex structural-acoustic systems." (provides* - capability)

Writer's note: Be specific and actionable. Name who benefits, how, and by how much: *"The proposed method can be used by calibration engineers to reduce measurement time by 40% without loss of accuracy."*

Key Discussion Connectors

Key finding: Engineering hedging modals (22.0/10k) are roughly one-third of the humanities rate (67.1/10k). Engineering Discussion sections are more assertive. The inferential markers *thus/therefore/hence* (20.6/10k) are at comparable density to the humanities.

Connector	Count	Per 10k	Function
may / might / could	1,341	22.0	Hedging modals
thus / therefore / hence / consequently	1,255	20.6	Inferential conclusion
although / though / while / whereas	1,041	17.1	Concessive contrast
however	918	15.1	Contrastive pivot
indeed / in fact	104	1.7	Emphatic reinforcement

The Discussion Move Cycle

For each major finding:

Restate (Move 1): *"The proposed model accurately predicts the load–displacement behaviour for all six configurations."*

Compare (Move 2): *"This is consistent with [12], who reported similar predictions for steel connections."*

Explain (Move 3): *"This is because the model captures the strain hardening effect that dominates the post-yield response."*

Then, after all findings are discussed:

Limitations (Move 4): *"A limitation of the current model is that it does not account for bolt hole clearance effects."*

Implications (Move 5): *"The model can be used by structural engineers to predict connection behaviour without requiring full nonlinear FE analysis."*

Implications for Writing

Explain Through Mechanisms

Use *this is because* + physical mechanism rather than *one possible explanation is* + theoretical framework.

Keep Limitations Brief and Technical

Name the specific technical limitation, state its consequence, and redirect to future work.

Foreground Practical Value

At 13.2/10k, practical implications are a major move. State who can use your findings, for what, and with what benefit.

Hedge Less Than You Think

Be appropriately cautious, but do not over-hedge clear findings.

Unit 9: Rhetorical Moves in the Conclusion

Engineering Corpus — 143 Conclusion sections, 48,544 words

Overview

The Conclusion compresses your paper into its essential claims: what was done, what was found, why it matters, and what comes next. In engineering, the Conclusion typically accounts for 8-10% of the paper.

The most distinctive feature is the emphasis on **contribution and novelty claims** (54.8/10k). A second distinctive feature is the **enumerated key findings** format (8.9/10k), where conclusions are presented as a numbered list.

Before You Write: Conceptualising Your Conclusion

Step 1: Reconnect with Your Gap

Re-read the niche from your Introduction. Write: *"Before this study, the problem was ___."*

Step 2: State Your Fill

Write: *"This paper has proposed/developed/demonstrated ___."* Check: does this address the gap?

Step 3: List Your Key Findings

Write 3-5 one-sentence findings. These may become the numbered conclusions.

Step 4: Identify Future Work

Name the specific experiment, extension, or parameter study that remains.

Move 1: Summarising Findings (41.2/10k)

Usage	Count	Per 10k	Function
in this paper/study/work	70	14.4	Framing the study
the results show/indicate/demonstrate	34	7.0	Data-driven summary
has been proposed/presented/developed	33	6.8	Passive method announcement
overall / in summary / to summarise	33	6.8	Summary marker
it was/has been shown/found/demonstrated	30	6.2	Passive finding restatement
Total Move 1	200	41.2	

Key finding: Move 1 density (41.2/10k) is quite high. The dominant opening is *in this paper/study/work* (14.4/10k). The passive method announcement *has been proposed/presented/developed* (6.8/10k) is distinctive of engineering.

Corpus examples:

> *"The six internal GEO assessment techniques introduced in this paper offer new and comprehensive insights into GEO collision likelihood."* (*in this paper* - framing + contribution)
> *"In this work a conceptual and computational methodology for decision support on waste heat recovery was developed and applied to an industrial case study."* (*in this work* - method + application)
> *"A new dataset has been presented which shows the potential influence that tool-holder diameter can have on the dynamics at the tool-tip."* (*has been presented* - novelty claim)

Writer's note: The opening sentence should capture the entire paper: *"In this paper, [method] has been [proposed/developed] for [purpose]."*

Move 2: Claiming Contribution and Novelty (54.8/10k)

This is the dominant move of the engineering Conclusion

Usage	Count	Per 10k	Function
novel / new / for the first time	99	20.4	Novelty claim
advantage / benefit / improvement	64	13.2	Advantage claim
the proposed [method/model/approach]	57	11.7	Method foregrounding
contribution / contributes to	29	6.0	Explicit contribution
superior / outperform / better than	17	3.5	Comparative superiority
Total Move 2	266	54.8	

Key finding: At 54.8/10k, the contribution and novelty language is very prominent. Engineering writers make strong, specific claims: *novel, new, for the first time* (20.4/10k); *advantage, improvement* (13.2/10k); *superior, outperforms* (3.5/10k).

Corpus examples:

> *"The proposed control method maintained satisfactory state estimation during handling manoeuvres, and achieved improved performance compared to standard LQ approaches." (the proposed + improved performance)*
> *"The proposed method prevents an EV from unnecessarily gaining kinetic energy if a part of it has to be dissipated while approaching the upcoming traffic signal." (the proposed method* + specific capability)
> *"This paper has proposed a novel method for manoeuvring ground vehicle tracking using a GMTI radar." (novel method* - explicit novelty claim)

Writer's note: Engineering Conclusions are assertive. State contributions directly with specific numbers: *"The proposed method achieves a 23% reduction in energy consumption."*

Move 3: Future Work (12.4/10k)

Usage	Count	Per 10k	Function
future work / future research / future studies	26	5.4	Naming future direction
can be extended / could be extended	25	5.1	Extension potential
should be [investigated/considered/explored]	7	1.4	Advisory recommendation
will be [extended/investigated/applied]	1	0.2	Planned work
Total Move 3	60	12.4	

Key finding: Future work language (12.4/10k) is present but not so prominent. Engineering future work statements are specific and technically oriented.

Corpus examples:

> *"Future work will provide more extensive experimental testing, especially in view of tool-life evaluation, and will further focus on exploring the use of solid lubricants."* (specific next steps)
> *"The future work includes system level analysis of the method through large scale simulations for different traffic scenarios."* (specific computational extension)
> *"However, future research is needed to investigate the statistical properties of randomness in probe error, and to refine error modelling."* (specific gap)

Writer's note: Be specific. Not *"Further research is needed"* but *"Future work will extend the model to include thermal effects and validate against field measurements at full scale."*

Move 4: Enumerated Key Findings (8.9/10k)

A distinctively engineering format: conclusions presented as a numbered list.

Usage	Count	Per 10k	Function
firstly/secondly/thirdly / (1) (2) (3)	27	5.6	Enumeration markers
can be summarised / can be drawn / as follows	10	2.1	List introduction
the main/key/principal findings/conclusions	6	1.2	Named findings
Total Move 4	43	8.9	

Key finding: The enumerated conclusion format is distinctively engineering. The pattern is: an introductory sentence followed by 3-6 numbered points, each stating a specific finding or claim.

Corpus examples:

> *"The main conclusions that can be drawn from this case study are: The decision support system is able to compare temporal availability of exergy between sources and sinks of waste heat energy."* (introductory sentence + list)
> *"Based on the experimental results, the following conclusions can be drawn: The tensile breakout capacity of headed anchors increases with increasing the thickness of concrete member." (based on* results + enumerated)

Writer's note: The enumerated format works well for 3-6 discrete findings. Each point should be one specific claim. If findings are interconnected, use text summary (Move 1) instead.

Engineering Conclusions

The Conclusion Template

Summarise (~30%): *"In this paper, [method] has been [proposed/developed] for [purpose]. The results show that [key finding]."*

Claim contribution (~30%): *"The proposed [method] achieves [specific performance]. Compared to [existing approaches], it offers [specific advantage]."*

Enumerate findings (optional, ~20%): *"The following conclusions can be drawn: (1)... (2)... (3)..."*

Future work (~20%): *"Future work will [specific extension]. [Specific experiment/validation] is also planned."*

Implications for Writing

Be Assertive

Contribution claims are prominent in the Conclusion. State what your method achieves, with specific numbers. This is not the place for hedging.

Use "The Proposed [Method]" as Subject

At 11.7/10k, this construction keeps your contribution in focus. It echoes the Introduction's *"this paper proposes..."* and creates a satisfying frame.

Consider the Enumerated Format

If you have 3-6 discrete findings, present them as a numbered list after an introductory sentence.

Be Specific About Future Work

Name the experiment, the extension, the parameter. Not *"More research is needed"* but *"Future work will extend the model to three-dimensional geometries and validate against field data."*

Unit 10: First and Last Sentences of Each Section

Engineering Corpus — 163 articles analysed

Overview

The first sentence of each section establishes what the section will do. The last sentence creates a bridge to the next section or closes the argument. These are often the most structurally important sentences in any paper.

The most striking finding is that engineering Introductions open with **broad contextual statements** (70%) rather than with the gap or purpose. Conclusions open with **purpose restatements** (42%) — *"In this paper, [method] has been [developed] for [purpose]"* — immediately recapturing the paper's core contribution.

Before You Look: Thinking About Openings and Closings

Opening: What does the reader need to know first? What frame should they bring to this section?

Closing: What should the reader carry forward into the next section? What is the takeaway?

Introduction

First Sentences (151 articles)

Key finding: Engineering Introductions overwhelmingly open with broad context (70%). This is consistent with the CARS model: Move 1 establishes territory before Move 2 identifies the niche.

Pattern	Frequency	Function
Broad contextual statement	70%	Establishing territory
Definition / context	18%	Defining a key concept
Gap / problem statement	6%	Jumping to the niche
Purpose / aim statement	4%	Announcing the study directly

Corpus examples:

> *"Energy is critical to all economic factors including adequate livelihood, shelter, and food for all life on earth."* (broad contextual — territory)
> *"Vibro-acoustic analysis is a necessary step for the virtual design of aerospace structures."* (definition/context — technology importance)
> *"Knowing the likelihood of collision for satellites is of extreme importance and interest to the space community."* (contextual + importance)

Writer's note: Do not open your Introduction with *"This paper proposes..."* Only 4% of engineering papers do this. Open by establishing why the topic matters.

Last Sentences (151 articles)

Pattern	Frequency	Function
Technical content (continuing)	58%	Paper flows into body
Finding preview / results summary	14%	Previewing outcomes
Purpose / aim statement	13%	Stating the study's purpose
Paper outline / section preview	11%	Mapping the paper structure

Key finding: The **paper outline** (11%) - *"Section 2 describes... Section 3 presents..."* - is characteristic of engineering.

> *"Section 6 shares the main conclusions of the proposed controller, its potential limitations and future work."* (paper outline)

"This paper presents a framework and an associated decision support tool specifically focused on waste heat recovery." (purpose statement)

Results

First Sentences (42 articles)

Pattern	Frequency	Function
General / contextual	67%	Setting up results context
Results / finding statement	17%	Jumping to findings
Method / approach reference	10%	Linking back to methods
Purpose statement	5%	Restating analysis goal

Key finding: Most Results sections open with context (67%): the test configuration, simulation setup, or parameter rather than jumping directly to a finding.

"All computations presented in this paper were performed using two HPC facilities." (context - computational setup)
"The output from the Monte Carlo simulations with 10,000 runs is shown in Figs. 13-17." (finding - direct figure reference)

Last Sentences (42 articles)

Pattern	Frequency	Function
Technical finding	57%	Reporting a final result
Table / figure reference	29%	Pointing to data display
Limitation / qualification	7%	Noting a constraint
Implication / significance	7%	Stating what the result means

Key finding: Results sections typically end mid-stream (57%) with a specific technical finding, not a summary or transition. The section simply stops when the last result has been reported. Table and fig ure references (29%) serve a similar function: the data display is the closing move.

Corpus examples:

> *"All three cases analysed are within the model validity, since the pressure term represents approximately 0.5% of the total thrust."* (technical finding - validation)
> *"These results facilitate the identification of the process bottlenecks in order to develop suitable direct applications."* (implication - significance

Discussion

First Sentences (19 articles)

Pattern	Frequency	Function
General / contextual	58%	Framing the discussion
Purpose / aim restatement	16%	Restating what was investigated
Results reference	11%	Linking back to findings
Definition / context	11%	Defining a discussion topic

Key finding: The most common opening frames the discussion, which is to follow.

Corpus examples:

> *"A number of aspects are worthy of further discussion."* (meta-comment - announcing the discussion)
> *"The primary advantage of the pruning heuristic as a global trajectory optimization method is clear from the results."* (results reference + evaluation)

Last Sentences (19 articles)

Pattern	Frequency	Function
Future work / extension	29%	Projecting next steps
Technical observation	24%	Continuing the analysis
Limitation / qualification	18%	Acknowledging constraints
Practical implication	18%	Suggesting applications
Contribution / significance	12%	Restating the contribution

Key finding: Discussion sections close more deliberately than Results sections. Future work (29%) and limitations (18%) are the characteristic closing moves. Both signal that the writer is stepping back from the findings to evaluate their scope. This contrasts with Results, which simply ends with the last data point.

Corpus examples:

> *"New methods need to be developed in the future studies to remove their effect on classification."* (future work)
> *"The authors consider this procedure a valid method for the extraction of areal surface texture information from XCT data, applicable to additively manufactured parts but with potential applications beyond the AM field."* (contribution - significance + extension)
> *"The values may have been erroneous and there may be other, more significant error sources not presented in this study."* (limitation - qualification)

Conclusion

Key finding: The **purpose restatement** (42%) e.g., *"In this paper, [method] has been [proposed/developed]"* is the most distinctively engineering Conclusion opener. It immediately recaptures the core contribution.

First Sentences (138 articles)

Pattern	Frequency	Function
Purpose / aim restatement	42%	"In this paper, [method] has been [developed]..."
General / contextual	44%	Contextual opening
Results / finding statement	6%	Jumping to key finding
Other	8%	

Corpus examples:

> *"This paper presents an evaluation of the Sound Pressure Level (SPL) of an aircraft cabin noise through Statistical Energy Analysis."* (purpose restatement)
> *"In this work a conceptual and computational methodology for decision support on waste heat recovery was developed and applied to an industrial case study."* (purpose restatement - method + application)
> *"Results indicate that a collision is likely to occur every 4 years for the entire GEO active satellite population against a 1 cm RSO catalogue."* (key finding - direct)

Last Sentences (138 articles)

Pattern	Frequency	Function
Other (technical content)	57%	Concluding technical detail
Future work	25%	Projecting next steps
Contribution restatement	18%	Closing with the contribution

Key finding: A quarter of engineering papers close with a **future work** sentence (25%). The alternative is a **contribution restatement** (18%).

Future work closers:

> *"Future research will focus on applying different intelligent control strategies to improve the positioning precision, trajectory tracking accu-*

racy and dynamic performance."
"Future work on the topic will include a thorough optimality assessment of the solutions derived in this work, as well as the extension of the continuous scheme to orbits of arbitrary eccentricity."

Contribution closers:

"A practical comparison between the proposed semi-active suspension system and fixed damping settings was also presented, demonstrating the effectiveness of the semi-active control method."
"Comprehensive numerical simulations on a realistic scenario are performed to demonstrate the advantages of the proposed method."

Summary: The Sentence Arc: Key Patterns

The Frame Effect

The Introduction opens broad and narrows (context → gap → purpose). The Conclusion opens specific and broadens (purpose → findings → future work). Together, they create a mirror: the Introduction is a funnel; the Conclusion is an inverted funnel.

The Paper Outline Sentence

The construction *"Section 2 describes... Section 3 presents..."* appears in 11% of engineering Introductions. This roadmap is characteristic of engineering and rare in humanities.

The "In This Paper" Opening

The construction *"In this paper, [method] has been [proposed/developed]"* is the dominant Conclusion opener (42%). It echoes the Introduction's promise and signals delivery.

Future Work as Final Word

One in four engineering papers ends with a future work sentence — positioning the research as part of an ongoing programme rather than a completed investigation.

Unit 11: Tense Distribution Across Sections

Engineering Corpus — Full corpus (163 articles, ~610,000 words)

Overview

Tense choice is quite complex. In general, we can say that it signals whether the writer is stating established knowledge, reporting completed procedures, describing current capabilities, or projecting future possibilities.

The dominant tense in engineering is the **present** (*is/are* at 255.7/10k). The present tense dominates because much of the paper describes systems, models, and relationships that hold in the present: *"The model is based on…"*, *"The system consists of…"*, *"Fig. 5 shows…"*.

Tense Density by Section (per 10,000 words)

The Tense Arc

Introduction: Present tense dominates (228.0/10k). Establishes or refers to "general truths," describes the current state of knowledge. Past tense may be used for citing prior work. Present perfect (19.7/10k) bridges past research to the present.

Body sections: Present tense remains dominant (235.3/10k) but past tense rises to its peak (88.7/10k) for procedural descriptions. Present tense is used to describe models, systems, and results.

Conclusion: Present tense drops to its lowest (184.7/10k). Present perfect peaks (24.2/10k) for *"has been proposed/developed/demonstrated"*. The modal *may* rises to 13.0/10k to indicate cautious projection.

Before You Check: Thinking About Tense as Meaning

A Quick Self-Diagnosis

• Am I using present tense for established facts and general truths?

• Am I using past tense for completed actions (procedures, experiments)?

• Am I using present tense for model descriptions, equations, and figure narration?

• Am I using present perfect for connecting past research to present relevance?

• Am I using modal verbs (*can, may, will*) for capabilities and projections?

Section-by-Section Tense Patterns

Introduction: Present Tense Dominates

Present tense for stating established knowledge:

> *"Cabin noise in passenger aircraft is one of the comfort parameters which creates straightaway discomfort when exceeding personal thresholds."* (general truth)

Present perfect for bridging the past to the present:

> *"Several methods have been proposed for noise prediction in aerospace structures."* (literature review)

Past tense for citing specific prior work:

> *"The incorporation of load disturbances into an LQ optimal control framework was proposed in Brezas and Smith (2013)."* (specific contribution)

Body Sections: Past tense for Procedures, Present tense for Descriptions

Past passive: reporting what was done:

> *"For the SC (scheme 2), the optimum volume was found 1.63 the minimum volume of the tank."* (completed procedure or finding)

Present passive: describing the model/system:

> *"An air/power block working fluid heat exchanger is used to transfer the heat from the heated air stream."* (system description)

Modal verb and passive: stating capabilities:

> *"Among these methods there is the Statistical Energy Analysis (SEA), which can be used both as theoretical and experimental technique."* (general capability)

Conclusion: Present Perfect for Summary, Modals for Projection

Present perfect: summarising the contribution:

> *"Finally, a financial model has been developed by MATLAB to compute the Levelized Cost of Energy (LCOE)." (has been developed)*

Modal verb: projecting future work:

> *"Future work will provide more extensive experimental testing." (will* - planned)
> *"The method can be extended to three-dimensional geometries." (can* - capability)

Key Findings

Engineering writing is far more present-tense dominant than writing in other disciplines. *Is/are* (255.7/10k) appears roughly three times more often than *was/were* (87.7/10k). This reflects models, equations, and figure narrations expressed as timeless truths.

The Past Passive Is Procedural

Past passive is concentrated in Methods sections. Engineering Results are often narrated in present tense (*"Fig. 5 shows that the displacement increases"*) because figures are treated as present realities.

Present Perfect Peaks in the Conclusion

Has/have been peaks at 24.2/10k in the Conclusion, nearly double the body rate. This reflects *"[Method] has been [proposed/developed/demonstrated]"*. Two-thirds of the instances are passive. The active uses are mostly *have shown, have presented, have proposed* etc.

Modal Verbs Are Distributed, Not Concentrated

Unlike some other duisciplines, engineering modal verbs s are more evenly distributed. The capability modal *can* (41.2/10k) is the dominant modal ,used for describing what methods can do, not for hedging.

I am...	Use...	Example
Stating a general truth	Present	"The system is governed by..."
Describing a model or figure	Present	"Eq. (3) yields...", "Fig. 5 shows..."
Reporting a completed procedure	Past passive	"The specimen was tested at..."
Citing specific prior work	Past	"[Author] proposed..."
Connecting past to present	Present perfect	"Several methods have been developed..."
Summarising your contribution	Present perfect	"A novel method has been proposed..."
Stating what the method can do	Modal can	"The method can be applied to..."
Projecting future work	Modal will	"Future work will extend..."
Hedging an uncertain claim	Modal may/might	"This may be due to..."

Unit 12: Narrating Figures, Tables, and Equations

Engineering Corpus — Full corpus (163 articles, ~610,000 words)

Overview

Engineering papers communicate through data displays far more than through text alone. The corpus contains approximately **53.1 display references per 10,000 words**: one reference every 19 words.

Display type	References	Per 10k	Share
Figures	2,567	42.2	79%
Tables	516	8.5	16%
Equations	149	2.4	5%
Total	3,232	53.1	100%

Figures dominate overwhelmingly (79%). Performance curves, contour plots, schematics, and comparison graphs are the primary vehicles for communicating results.

Narrating Figures (42.2/10k)

Key finding: The five constructions offer different rhetorical effects. *Fig. N shows* introduces a new figure. *X is shown in Fig. N* foregrounds the finding. *As shown in Fig. N* is a back-reference. *See Fig. N* is a directive. *It can be seen* is impersonal. It is used frequently in engineering. Make sure that, in fact, “it can be seen”.

Construction	Count	Per 10k	Function
Fig. N shows/presents/illustrates	350	5.7	Figure as subject (active)
as shown/seen/illustrated in Fig. N	260	4.3	Parenthetical back-reference
X is/are shown/plotted in Fig. N	254	4.2	Finding as subject (passive)
see/from Fig. N	166	2.7	Directive or locative reference
it can be seen/observed from	46	0.8	Impersonal observation

Corpus examples:

> *"Fig. 9 shows the SPL evaluated at 'Head Cavities' for some section and for all the 5 investigated configurations." (Fig. N shows* - introducing)
> *"The output from the Monte Carlo simulations with 10,000 runs is shown in Figs. 13-17." (X is shown in* - finding as topic)
> *"From Fig. 5, it can be seen that the traffic signal state ID starts with 1, i.e. state green."* (impersonal observation)

The Figure Narration Sequence

1. Introduce: "Fig. 5 shows the variation of X with Y for all three configurations."

2. Direct attention: "The maximum displacement occurs at the midspan."

3. Quantify: "Configuration A shows a peak of 12.3 mm, compared to 8.7 mm for B."

4. Comment: "This reduction is consistent with the increased stiffness."

Narrating Tables (8.5/10k)

Construction	Count	Per 10k	Function
X is/are shown/listed/given in Table N	103	1.7	Finding as subject
Table N shows/presents/lists	88	1.4	Table as subject
as shown/given/listed in Table N	26	0.4	Parenthetical back-reference
see Table N	25	0.4	Directive reference

Key finding: Unlike figures, table narration slightly favours the finding-as-subject construction (1.7/10k vs. 1.4/10k). Tables contain multiple data points, and so writers tend to foreground specific data.

Corpus examples:

> *"Table 3 shows the electric vehicle's energy consumption with and without the proposed ECC." (Table N shows)*
> *"Table 8 shows that the tracking accuracy on a 50 μm-diameter circular trajectory was less than 5 nm." (Table N shows that* - specific finding)
> *"Table 2 lists the computational details including grid sizes for the different computations." (Table N lists* - listing)

Narrating Equations

Construction	Count	Per 10k	Function
substituting / combining / rearranging	233	3.8	Derivation step
is given by / is defined as / is expressed as	130	2.1	Equation introduction
where X is/are/denotes	75	1.2	Variable definition
can be written as / can be expressed as	24	0.4	Alternative formulation

Key finding: The dominant equation narration pattern is the **derivation step** (*substituting, combining, rearranging* at 3.8/10k), not the equation reference. The introduction *is given by* (2.1/10k) presents equations for the first time. The variable definition *where X is/ denotes* (1.2/10k) always follows immediately.

Corpus examples:

> *"The equation of motion is expressed as [equation] considering the boundary conditions for the system." (is expressed as* - introduction)
> *"This force can be expressed as a function of the slot's position and width." (can be expressed as* - formulation)
> *"The fin axis system is defined as x positive forward, y positive outboard." (is defined as* - coordinate definition)

The Equation Narration Sequence

1. Introduce: *"The governing equation is given by"* [equation]

2. Define: *"where ρ is the density, u is the velocity, and p is the pressure."*

3. Derive: *"Substituting Eq. (3) into Eq. (1) and rearranging yields"* [equation]

4. State result: *"The resulting expression shows that X depends linearly on Y."*

Implications for Writing

Plan Your Figures First

At 42.2/10k, figure references are among the densest constructions. Plan figures before writing the text; the text serves the figures.

Vary Your Display Reference Constructions

Cycle through the five figure narration constructions. Introduce with *Fig. N shows*, then use *as shown in Fig. N* for subsequent references.

Always Narrate; Never Assume

Every figure, table, and equation should be explicitly introduced in the text before it appears.

Follow Each Equation with Variable Definitions

The *where X is/denotes* construction (1.2/10k) should immediately follow every equation that introduces new variables.

Unit 13: Citation Distribution and Integration

Engineering Corpus — Full corpus (163 articles, ~610,000 words)

Overview

Engineering papers cite by number, not by name. Numeric citations account for **148.9/10k**, roughly one citation every 67 words. Author–year citations appear at just 1.1/10k.

Citation Density by Section

Section	Citations	Per 10k	Character
Introduction	2,351	112.2	Heaviest — one every 89 words
Body (Methods + Results)	2,137	64.8	Moderate — one every 154 words
Conclusion	74	16.3	Sparse — one every 614 words

Key finding: The Introduction carries more than half the paper's citations, reflecting the CARS model. The Conclusion is nearly citation-free, focusing on what *the* paper contributes.

The Citation Arc

Introduction: Dense: establishing territory, reviewing literature, identifying the gap.

Body: Moderate: referencing methods, standards, and comparison data.

Conclusion: Minimal: looking forward (contribution, future work), not backward.

Citation Integration Patterns

Key finding: Engineering citations are overwhelmingly **non-integral**, that is, the reference number appears at the end of a clause. The integral form (*[12] showed that...* at 2.9/10k) is used only when the cited work's specific finding needs to be foregrounded.

Corpus examples:

> *"Several studies can be found in literature [12,13]." (non-integral -* at end of clause)
> *"Similar to [1], the cost function does not aim to minimize acceleration." (integral* - reference as topic)

Implications for Writing

Front-Load Your Citations

Introduction should be citation-dense (112.2/10k). Every claim about the state of the field is likely to need a reference.

Use Non-Integral Citation by Default

Place reference numbers at end of clauses. Reserve integral citation for critical findings.

Let the Conclusion Speak for Itself

The Conclusion is nearly citation-free (16.3/10k). Claim *your* contribution.

Unit 14: Hedging and Boosting

Engineering Corpus — Full corpus (163 articles, ~610,000 words)

Overview

The most striking finding is that the **Conclusion has the highest density of both hedging and boosting**: hedging at 58.0/10k and boosting at 62.4/10k. This reflects the Conclusion's dual function: cautious projection and strong contribution claims.

Hedging and Boosting by Section

Section	Hedging	Boosting
Introduction	39.2/10k	35.1/10k
Body	47.7/10k	37.4/10k
Conclusion	58.0/10k	62.4/10k

Key finding: Engineering's highest hedging section (Conclusion 58.0/10k) is still below that of other disciplines. Engineering writing tends to be more assertive overall, supported by empirical data.

The Hedging Vocabulary

Key finding: The dominant "hedge" *can* (41.2/10k) refers primarily to capability rather than possibility. True epistemic hedges are *may* (11.1/10k) and *could* (8.9/10k).

Hedge	Count	Per 10k	Function
can	2,511	41.2	Capability (not epistemic)
may	673	11.1	Epistemic possibility
could	541	8.9	Conditional possibility
possible/possibly	420	6.9	Named possibility
should	410	6.7	Recommendation
would	402	6.6	Conditional/hypothetical
relatively/somewhat/slightly	322	5.3	Degree softener
suggest(s/ed)	292	4.8	Tentative reporting
approximately/roughly	224	3.7	Numerical approximation
might	127	2.1	Weak possibility

The Boosting Vocabulary

Booster	Count	Per 10k	Function
significant(ly)	677	11.1	Marking significance
highly/very/extremely	648	10.6	Intensifier
important/crucial/critical	636	10.4	Importance claim
demonstrate(s/d)/prove(s/d)	481	7.9	Evidence verb
novel/unique/innovative	204	3.3	Novelty claim
clearly/obviously	107	1.8	Epistemic certainty
indeed / in fact	104	1.7	Emphatic reinforcement

Hedge Explanations, Not Validated Results

If validated, state results directly. Hedge only when speculating about causes.

Use "Can" for Capability, "May" for Uncertainty

The most frequent pattern is: *Can* = what is technically possible. *May* = what is epistemically uncertain. This is not a hard and fast rule.

Boost Your Contribution

Engineering Conclusions boost at 62.4/10k. Use *novel, significant, demonstrate, improve* to focus on your contribution.

Unit 15: Authorial Presence

Engineering Corpus — Full corpus (163 articles, ~610,000 words)

Overview

Engineering writers maintain authorial presence through **constructions** that keep the work or method as the grammatical subject. *This paper/study/work* (15.2/10k) appears more frequently than *we* (10.8/10k).

Authorial Presence Vocabulary

Construction	Count	Per 10k	Function
this paper / this study / this work	923	15.2	Paper as subject
we (subject pronoun)	657	10.8	First-person agency
the proposed [method/model]	255	4.2	Method as proxy
our (possessive)	204	3.3	First-person ownership
it is [adj] to / it should be noted	173	2.8	Impersonal evaluation
the author(s)	137	2.2	Third-person self-reference
the present [study/work]	69	1.1	Formal self-reference

Authorial Presence by Section

Section	we	our	this paper/study	the proposed
Introduction	7.7/10k	1.9/10k	16.0/10k	3.6/10k
Body	6.5/10k	2.1/10k	7.9/10k	3.1/10k
Conclusion	10.1/10k	2.4/10k	27.3/10k	11.9/10k

Key finding: Authorial presence peaks in the Conclusion: *this paper/study* at 27.3/10k, *the proposed* at 11.9/10k. The Body has lowest presence; the methods and results are allowed to speak for themselves.

Three Modes of Self-Reference

Mode 1: First Person (we/our: 14.1/10k)

Used for decisions and claims:

> *"We propose a practical limited feedback approach."*

Mode 2: Paper as Subject (this paper/study: 15.2/10k)

Used for framing purpose and contribution:

> *"This paper presents a framework specifically focused on waste heat recovery."*

Mode 3: Method as Proxy (the proposed [method]: 4.2/10k)

Distinctive in engineering articles. The contribution is the main focus:

> *"The proposed control method maintained satisfactory state estimation."*

Implications for Writing

Use "This Paper" for Framing, "We" for Decisions

Both are appropriate; use each for its intended function.

Use "The Proposed [Method]" for Contribution Claims

Keeps focus on the technical contribution: *"The proposed model achieves good agreement."*

Match Your Section's Expectations

Minimal authorial presence in Body. Increase in Introduction (framing) and Conclusion (claiming).

Unit 16: The Paper as Argument

Engineering Corpus — Capstone Unit

Overview

Every engineering research article makes a single macro-argument: **There is a problem, and we solve it.** The entire paper serves this argument. This capstone unit synthesises Units 1-15.

The macro-argument has four stages: (1) The problem exists. (2) Here is our solution. (3) The solution works. (4) Here is what it means.

Stage 1: The Problem Exists (Introduction)

Feature	Density	Function
Citations	112.2/10k	Others have worked on this
Importance claims	High	The topic matters
Gap language (however, limited)	~15/10k	What is missing
this paper proposes/presents	16.0/10k	Announcing the solution
Present tense (is/are)	228.0/10k	General truths
Present perfect (has been)	19.7/10k	Past work → present need

Logic: *Others have worked on X [citations]. X is important [boosting]. But Y remains unsolved [gap]. This paper proposes Z [announcement].*

Stage 2: Here Is Our Solution (Methods/Theory)

Feature	Density	Function
Past passive (was/were)	88.7/10k (peak)	Completed procedures
is given by	2.1/10k	Mathematical framework
where X is/denotes	1.2/10k	Defining components
substituting/combining	3.8/10k	Building the case
Citations	64.8/10k	Referencing methods

Stage 3: The Solution Works (Results + Validation)

Feature	Density	Function
Figure references	42.2/10k	Showing evidence
Directional change	54.4/10k	What happened
Agreement language	1.5/10k	Judging model quality
compared with/to	9.3/10k	Comparing with reference
Contrastive however	18.7/10k	Comparing conditions

Stage 4: Here Is What It Means (Disc + Conc)

Feature	Density	Function
this is because	1.3/10k	Explaining mechanisms
enables/allows/provides	8.0/10k	Practical value
Contribution claims (novel)	54.8/10k (Conc.)	What is new
the proposed [method]	11.9/10k (Conc.)	Foregrounding contribution
Present perfect (has been)	24.2/10k (Conc.)	Summarising
Future work	5.4/10k (Conc.)	Projecting what comes next

The Cross-Section Arcs

Citation Arc

Introduction (112.2) → Body (64.8) → Conclusion (16.3). Engage with others at the start; claim your own contribution at the end. The abstract is typically citation-free

Tense Arc

Present dominates; past peaks in Body (88.7/10k); present perfect peaks in Conclusion (24.2/10k). The abstract has its own profile. Check Unit 2.

Hedging Arc

Introduction (39.2) → Body (47.7) → Conclusion (58.0). Hedging *increases*: from established knowledge to cautious projection.

Boosting Arc

Introduction (35.1) → Body (37.4) → Conclusion (62.4). Boosting peaks sharply in the Conclusion.

Authorial Presence Arc

Introduction (16.0) → Body (7.9) → Conclusion (27.3). Most visible at beginning and end.

The Macro-Argument

Abstract: *"Here is the problem, the approach and the key result in one paragraph."*

Introduction: *"There is a problem. Others have tried, but gaps remain. This paper proposes a solution."*

Methods: *"Here is how the solution works, in precise detail."*

Results + Validation: *"Here is evidence that the solution works."*

Discussion: *"Here is why it works, how it compares, and what it means."*

Conclusion: "The solution is novel and effective. Future work will extend it."

A Final Checklist

1. Does the Introduction establish a clear gap? (Unit 3)

2. Are your assumptions stated explicitly? (Unit 4)

3. Can a reader reproduce your method? (Unit 4)

4. Do your figures tell the story? (Units 6, 12)

5. Have you validated against independent data? (Unit 7)

6. Have you explained results through mechanisms? (Unit 8)

7. Does the Conclusion claim contribution strongly? (Unit 9)

8. Do first and last sentences frame each section? (Unit 10)

9. Is your tense consistent within each function? (Unit 11)

10. Are your citations front-loaded? (Unit 13)

11. Have you hedged interpretations but boosted contributions? (Unit 14)

12. Does the Conclusion answer the Introduction's question? (Units 3 and 9)

13. Does the Abstract state the gap, method and key finding? (Unit 2)

Hsppy Writing!

www.ingramcontent.com/pod-product-compliance
Lightning Source LLC
LaVergne TN
LVHW050936210926
848482LV00013B/639
9780940753419